Cloud Application Architectures

Cloud Application Architectures

George Reese

Cloud Application Architectures
by George Reese

Printed in the United States of America.

Published by O'Reilly Media, Inc., 1005 Gravenstein Highway North, Sebastopol, CA 95472.

O'Reilly books may be purchased for educational, business, or sales promotional use. Online editions are also available for most titles (*http://safari.oreilly.com*). For more information, contact our corporate/institutional sales department: (800) 998-9938 or *corporate@oreilly.com*.

Editor: Andy Oram
Production Editor: Sumita Mukherji
Copyeditor: Genevieve d'Entremont
Proofreader: Kiel Van Horn
Indexer: Joe Wizda
Cover Designer: Mark Paglietti
Interior Designer: David Futato
Illustrator: Robert Romano

Printing History:
April 2009: First Edition.

ISBN: 978-0-596-15636-7

[V] [9/09]

1250274173

CONTENTS

PREFACE

In 2003, I jumped off the entrepreneurial cliff and started the company Valtira. In a gross oversimplification, Valtira serves the marketing function for companies in much the same way that SalesForce.com serves the sales function. It does online campaign management, customer relationship management (CRM) integration with marketing programs, personalized web content, and a lot of other marketing things. Valtira's business model differed in one key way from the SalesForce.com business model: the platform required you to build your website on top of the content management system (CMS) at its core.

This CMS requirement made Valtira much more powerful than its competition as a Software as a Service (SaaS) marketing tool. Unfortunately, it also created a huge barrier to entry for Valtira solutions. While many companies end up doing expensive CRM integration services engagements with SalesForce.com, you can get started on their platform without committing to a big integration project. Valtira, on the other hand, demanded a big web development project of each customer.

In 2007, we decided to alter the equation and began making components of the Valtira platform available *on-demand*. In other words, we changed our software so marketers could register via the Valtira website and immediately begin building landing pages or developing personalized widgets to stick on their websites.

Our on-demand application had a different risk profile than the other deployments we managed. When a customer built their website on top of the Valtira Online Marketing Platform, they selected the infrastructure to meet their availability needs and paid for that infrastructure.

If they had high-availability needs, they paid for a high-availability managed services environment at ipHouse or Rackspace and deployed our software into that infrastructure. If they did not have high-availability needs, we provided them with a shared server infrastructure that they could leverage.

The on-demand profile is different—everyone always expects an on-demand service to be available, regardless of what they are paying for it. I priced out the purchase of a starter high-availability environment for deploying the Valtira platform that consisted of the following components:

- A high-end load balancer
- Two high-RAM application servers
- Two fast-disk database servers
- Assorted firewalls and switches
- An additional half-rack with our ISP

Did I mention that Valtira is entirely self-funded? Bank loans, management contributions, and starter capital from family is all the money we have ever raised. Everything else has come from operational revenues. We have used extra cash to grow the business and avoided any extravagances. We have always managed our cash flow very carefully and were not excited about the prospect of this size of capital expense.

I began looking at alternatives to building out my own infrastructure and priced out a managed services infrastructure with several providers. Although the up-front costs were modest enough to stomach, the ongoing costs were way too high until we reached a certain level of sales. That's when I started playing with Amazon Web Services (AWS).

AWS promised us the ability to get into a relatively high-availability environment that roughly mirrored our desired configuration with no up-front cash and a monthly expense of under $1,000. I was initially very skeptical about the whole thing. It basically seemed too good to be true. But I started researching....

That's the first thing you should know about the cloud: "But I started researching." If you wanted to see whether your application will work properly behind a high-end load balancer across two application servers, would you ever go buy them just to see if it would work out OK? I am guessing the answer to that question is no. In other words, even if this story ended with me determining that the cloud was not right for Valtira's business needs, the value of the cloud is already immediately apparent in the phrase, "But I started researching."

And I encountered problems. First, I discovered how the Amazon cloud manages IP addresses. Amazon assigns all addresses dynamically, you do not receive any netblocks, and—at that *time—there* was no option for static IP address assignment. We spent a small amount of time on this challenge and figured we could craft an automated solution to this issue. My team moved on to the next problem.

Our next challenge was Amazon's lack of persistent storage. As with the issue of no static IP addresses, this concern no longer exists. But before Amazon introduced its Elastic Block Storage services, you lost all your data if your EC2 instance went down. If Valtira were a big company with a lot of cash, we would have considered this a deal-breaker and looked elsewhere.

We almost did stop there. After all, the Valtira platform is a database-driven application that cannot afford any data loss. We created a solution that essentially kept our MySQL slave synced with Amazon S3 (which was good enough for this particular use of the Valtira platform) and realized this solution had the virtue of providing automated disaster recovery.

This experimentation continued. We would run into items we felt were potential deal-breakers only to find that we could either develop a workaround or that they actually encouraged us to do things a better way. Eventually, we found that we could make it all work in the Amazon cloud. We also ended up spinning off the tools we built during this process into a separate company, enStratus.

Today, I spend most of my time moving other companies into the cloud on top of the enStratus software. My customers tend to be more concerned with many of the security and privacy aspects of the cloud than your average early-adopter. The purpose of this book is to help you make the transition and prepare your web applications to succeed in the cloud.

Audience for This Book

I have written this book for technologists at all career levels. Whether you are a developer who needs to write code for the cloud, or an architect who needs to design a system for the cloud, or an IT manager responsible for the move into the cloud, you should find this book useful as you prepare your journey.

This book does not have a ton of code, but here and there I have provided examples of the way I do things. I program mostly in Java and Python against MySQL and the occasional SQL Server or Oracle database. Instead of providing a bunch of Java code, I wanted to provide best practices that fit any programming language.

If you design, build, or maintain web applications that might be deployed into the cloud, this book is for you.

Organization of the Material

The first chapter of this book is for a universal audience. It describes what I mean by "the cloud" and why it has value to an organization. I wrote it at such a level that your CFO should be able to read the chapter and understand why the cloud is so useful.

In the second chapter, I take a bit of a diversion and provide a tutorial for the Amazon cloud. The purpose of this book is to provide best practices that are independent of whatever cloud you are using. My experience, however, is mostly with the Amazon cloud, and the Amazon

Web Services offerings make up the bulk of the market today. As a result, I thought it was critical to give the reader a way to quickly get started with the Amazon cloud as well as a common ground for discussing terms later in the book.

If you are interested in other clouds, I had help from some friends at Rackspace and GoGrid. Eric "E. J." Johnson from Rackspace has reviewed the book for issues that might be incompatible with their offering, and Randy Bias from GoGrid has done the same for their cloud infrastructure. Both have provided appendixes that address the specifics of their company offerings.

Chapter 3 prepares you for the cloud. It covers what you need to do and how to analyze the case for the move into the cloud.

Chapters 4 through 7 dive into the details of building web applications for the cloud. Chapter 4 begins the move into the cloud with a look at transactional web application architectures and how they need to change in the cloud. Chapter 5 confronts the security concerns of cloud computing. Chapter 6 shows how the cloud helps you better prepare for disaster recovery and how you can leverage the cloud to drive faster recoveries. Finally, in Chapter 7, we address how the cloud changes perspectives on application scaling—including automated scaling of web applications.

Conventions Used in This Book

The following typographical conventions are used in this book:

Italic
: Indicates new terms, URLs, filenames, Unix utilities, and command-line options.

`Constant width`
: Indicates the contents of files, the output from commands, and generally anything found in programs.

`Constant width bold`
: Shows commands or other text that should be typed literally by the user, and parts of code or files highlighted for discussion.

`Constant width italic`
: Shows text that should be replaced with user-supplied values.

Using Code Examples

This book is here to help you get your job done. In general, you may use the code in this book in your programs and documentation. You do not need to contact us for permission unless you're reproducing a significant portion of the code. For example, writing a program that uses several chunks of code from this book does not require permission. Selling or distributing a CD-ROM of examples from O'Reilly books does require permission. Answering a question by citing this book and quoting example code does not require permission. Incorporating a significant amount of example code from this book into your product's documentation does require permission.

We appreciate, but do not require, attribution. An attribution usually includes the title, author, publisher, and ISBN. For example, "*Cloud Application Architectures* by George Reese. Copyright 2009 George Reese, 978-0-596-15636-7."

If you feel your use of code examples falls outside fair use or the permission given above, feel free to contact us at *permissions@oreilly.com.*

Safari® Books Online

Safari Books Online

When you see a Safari® Books Online icon on the cover of your favorite technology book, that means the book is available online through the O'Reilly Network Safari Bookshelf.

Safari offers a solution that's better than e-books. It's a virtual library that lets you easily search thousands of top tech books, cut and paste code samples, download chapters, and find quick answers when you need the most accurate, current information. Try it for free at *http://my .safaribooksonline.com.*

We'd Like Your Feedback!

We at O'Reilly have tested and verified the information in this book to the best of our ability, but mistakes and oversights do occur. Please let us know about errors you may find, as well as your suggestions for future editions, by writing to:

O'Reilly Media, Inc.
1005 Gravenstein Highway North
Sebastopol, CA 95472
800-998-9938 (in the U.S. or Canada)
707-829-0515 (international or local)
707-829-0104 (fax)

We have a web page for the book where we list errata, examples, or any additional information. You can access this page at:

http://www.oreilly.com/catalog/9780596156367

To comment or ask technical questions about this book, send email to:

bookquestions@oreilly.com

For more information about our books, conferences, software, Resource Centers, and the O'Reilly Network, see our website at:

http://www.oreilly.com

Acknowledgments

This book covers so many disciplines and so many technologies, it would have been impossible for me to write it on my own.

First, I would like to acknowledge the tremendous help I received from Randy Bias at GoGrid and E. J. Johnson at Rackspace. My experience in cloud infrastructure has been entirely with Amazon Web Services, and Randy and E. J. spent a significant amount of time reviewing the book for places where the discussion was specific to AWS. They also wrote the appendixes on the GoGrid and Rackspace offerings.

Next, I would like to thank everyone who read each chapter and provided detailed comments: John Allspaw, Jeff Barr, Christofer Hoff, Theo Schlossnagle, and James Urquhart. They each brought very unique expertise into the technical review of this book, and the book is much better than it otherwise would have been, thanks to their critical eyes.

In addition, a number of people have reviewed and provided feedback on selected parts of the book: David Bagley, Morgan Catlin, Mike Horwath, Monique Reese, Stacey Roelofs, and John Viega.

Finally, I owe the most thanks on this book to Andy Oram and Isabel Kunkle from O'Reilly. I have said this in other places, but I need to say it here: their editing makes me a better writer.

CHAPTER ONE

Cloud Computing

THE HALLMARK OF ANY BUZZWORD is its ability to convey the appearance of meaning without conveying actual meaning. To many people, the term *cloud computing* has the feel of a buzzword.

It's used in many discordant contexts, often referencing apparently distinct things. In one conversation, people are talking about Google Gmail; in the next, they are talking about Amazon Elastic Compute Cloud (at least it has "cloud" in its name!).

But cloud computing is not a buzzword any more than the term *the Web* is. Cloud computing is the evolution of a variety of technologies that have come together to alter an organization's approach to building out an IT infrastructure. Like the Web a little over a decade ago, there is nothing fundamentally new in any of the technologies that make up cloud computing. Many of the technologies that made up the Web existed for decades when Netscape came along and made them accessible; similarly, most of the technologies that make up cloud computing have been around for ages. It just took Amazon to make them all accessible to the masses.

The purpose of this book is to empower developers of transactional web applications to leverage cloud infrastructure in the deployment of their applications. This book therefore focuses on the cloud as it relates to clouds such as Amazon EC2, more so than Google Gmail. Nevertheless, we should start things off by setting a common framework for the discussion of cloud computing.

The Cloud

The cloud is not simply the latest fashionable term for the Internet. Though the Internet is a necessary foundation for the cloud, the cloud is something more than the Internet. The cloud is where you go to use technology when you need it, for as long as you need it, and not a minute more. You do not install anything on your desktop, and you do not pay for the technology when you are not using it.

The cloud can be both software and infrastructure. It can be an application you access through the Web or a server that you provision exactly when you need it. Whether a service is software or hardware, the following is a simple test to determine whether that service is a cloud service:

> If you can walk into any library or Internet cafe and sit down at any computer without preference for operating system or browser and access a service, that service is cloud-based.

I have defined three criteria I use in discussions on whether a particular service is a cloud service:

- The service is accessible via a web browser (nonproprietary) or web services API.
- Zero capital expenditure is necessary to get started.
- You pay only for what you use as you use it.

I don't expect those three criteria to end the discussion, but they provide a solid basis for discussion and reflect how I view cloud services in this book.

If you don't like my boiled-down cloud computing definition, James Governor has an excellent blog entry on "15 Ways to Tell It's Not Cloud Computing," at *http://www.redmonk.com/jgovernor/2008/03/13/15-ways-to-tell-its-not-cloud-computing*.

Software

As I mentioned earlier, cloud services break down into software services and infrastructure services. In terms of maturity, software in the cloud is much more evolved than hardware in the cloud.

Software as a Service (SaaS) is basically a term that refers to software in the cloud. Although not all SaaS systems are cloud systems, most of them are.

SaaS is a web-based software deployment model that makes the software available entirely through a web browser. As a user of SaaS software, you don't care where the software is hosted, what kind of operating system it uses, or whether it is written in PHP, Java, or .NET. And, above all else, you don't have to install a single piece of software anywhere.

Gmail, for example, is nothing more than an email program you use in a browser. It provides the same functionality as Apple Mail or Outlook, but without the fat client. Even if your domain does not receive email through Gmail, you can still use Gmail to access your mail.

SalesForce.com is another variant on SaaS. SalesForce.com is an enterprise customer relationship management (CRM) system that enables sales people to track their prospects and leads, see where those individuals sit in the organization's sales process, and manage the workflow of sales from first contact through completion of a sale and beyond. As with Gmail, you don't need any software to access SalesForce.com: point your web browser to the SalesForce.com website, sign up for an account, and get started.

SaaS systems have a few defining characteristics:

Availability via a web browser
: SaaS software never requires the installation of software on your laptop or desktop. You access it through a web browser using open standards or a ubiquitous browser plug-in. Cloud computing and proprietary desktop software simply don't mix.

On-demand availability
: You should not have to go through a sales process to gain access to SaaS-based software. Once you have access, you should be able to go back into the software any time, from anywhere.

Payment terms based on usage
: SaaS does not need any infrastructure investment or fancy setup, so you should not have to pay any massive setup fees. You should simply pay for the parts of the service you use as you use them. When you no longer need those services, you simply stop paying.

Minimal IT demands
: If you don't have any servers to buy or any network to build out, why do you need an IT infrastructure? While SaaS systems may require some minimal technical knowledge for their configuration (such as DNS management for Google Apps), this knowledge lays within the realm of the power user and not the seasoned IT administrator.

One feature of some SaaS deployments that I have intentionally omitted is multitenancy. A number of SaaS vendors boast about their multitenancy capabilities—some even imply that multitenancy is a requirement of any SaaS system.

A multitenant application is server-based software that supports the deployment of multiple clients in a single software instance. This capability has obvious advantages for the SaaS vendor that, in some form, trickle down to the end user:

- Support for more clients on fewer hardware components
- Quicker and simpler rollouts of application updates and security patches
- Architecture that is generally more sound

The ultimate benefit to the end user comes indirectly in the form of lower service fees, quicker access to new functionality, and (sometimes) quicker protection against security holes. However, because a core principle of cloud computing is a lack of concern for the underlying architecture of the applications you are using, the importance of multitenancy is diminished when looking at things from that perspective.

As we discuss in the next section, virtualization technologies essentially render the architectural advantages of multitenancy moot.

Hardware

In general, hardware in the cloud is conceptually harder for people to accept than software in the cloud. Hardware is something you can touch: you own it; you don't license it. If your server catches on fire, that disaster matters to you. It's hard for many people to imagine giving up the ability to touch and own their hardware.

With hardware in the cloud, you request a new "server" when you need it. It is ready as quickly as 10 minutes after your request. When you are done with it, you release it and it disappears back into the cloud. You have no idea what physical server your cloud-based server is running, and you probably don't even know its specific geographic location.

THE BARRIER OF OLD EXPECTATIONS

The hardest part for me as a vendor of cloud-based computing services is answering the question, "Where are our servers?" The real answer is, inevitably, "I don't know—somewhere on the East Coast of the U.S. or Western Europe," which makes some customers very uncomfortable. This lack of knowledge of your servers' location, however, provides an interesting physical security benefit, as it becomes nearly impossible for a motivated attacker to use a physical attack vector to compromise your systems.

The advantages of a cloud infrastructure

Think about all of the things you have to worry about when you own and operate your own servers:

Running out of capacity?
: Capacity planning is always important. When you own your own hardware, however, you have two problems that the cloud simplifies for you: what happens when you are wrong (either overoptimistic or pessimistic), and what happens if you don't have the expansion capital when the time comes to buy new hardware. When you manage your own infrastructure, you have to cough up a lot of cash for every new Storage Area Network (SAN) or every new server you buy. You also have a significant lead time from the moment you decide to make a purchase to getting it through the procurement process, to taking delivery, and finally to having the system racked, installed, and tested.

What happens when there is a problem?
: Sure, any good server has redundancies in place to survive typical hardware problems. Even if you have an extra hard drive on hand when one of the drives in your RAID array

fails, someone has to remove the old drive from the server, manage the RMA,* and put the new drive into the server. That takes time and skill, and it all needs to happen in a timely fashion to prevent a complete failure of the server.

What happens when there is a disaster?
: If an entire server goes down, unless you are in a high-availability infrastructure, you have a disaster on your hands and your team needs to rush to address the situation. Hopefully, you have solid backups in place and a strong disaster recovery plan to get things operational ASAP. This process is almost certainly manual.

Don't need that server anymore?
: Perhaps your capacity needs are not what they used to be, or perhaps the time has come to decommission a fully depreciated server. What do you do with that old server? Even if you give it away, someone has to take the time to do something with that server. And if the server is not fully depreciated, you are incurring company expenses against a machine that is not doing anything for your business.

What about real estate and electricity?
: When you run your own infrastructure (or even if you have a rack at an ISP), you may be paying for real estate and electricity that are largely unused. That's a very ungreen thing, and it is a huge waste of money.

None of these issues are concerns with a proper cloud infrastructure:

- You add capacity into a cloud infrastructure the minute you need it, and not a moment sooner. You don't have any capital expense associated with the allocation, so you don't have to worry about the timing of capacity needs with budget needs. Finally, you can be up and running with new capacity in minutes, and thus look good even when you get caught with your pants down.
- You don't worry about any of the underlying hardware, ever. You may never even know if the physical server you have been running on fails completely. And, with the right tools, you can automatically recover from the most significant disasters while your team is asleep.
- When you no longer need the same capacity or you need to move to a different virtual hardware configuration, you simply deprovision your server. You do not need to dispose of the asset or worry about its environmental impact.
- You don't have to pay for a lot of real estate and electricity you never use. Because you are using a fractional portion of a much beefier piece of hardware than you need, you are maximizing the efficiency of the physical space required to support your computing needs. Furthermore, you are not paying for an entire rack of servers with mostly idle CPU cycles consuming electricity.

* Return merchandise authorization. When you need to return a defective part, you generally have to go through some vendor process for returning that part and obtaining a replacement.

Hardware virtualization

Hardware virtualization is the enabling technology behind many of the cloud infrastructure vendors offerings, including Amazon Web Services (AWS).† If you own a Mac and run Windows or Linux inside Parallels or Fusion, you are using a similar virtualization technology to those that support cloud computing. Through virtualization, an IT admin can partition a single physical server into any number of virtual servers running their own operating systems in their allocated memory, CPU, and disk footprints. Some virtualization technologies even enable you to move one running instance of a virtual server from one physical server to another. From the perspective of any user or application on the virtual server, no indication exists to suggest the server is not a real, physical server.

A number of virtualization technologies on the market take different approaches to the problem of virtualization. The Amazon solution is an extension of the popular open source virtualization system called Xen. Xen provides a hypervisor layer on which one or more guest operating systems operate. The hypervisor creates a hardware abstraction that enables the operating systems to share the resources of the physical server without being able to directly access those resources or their use by another guest operating system.

A common knock against virtualization—especially for those who have experienced it in desktop software—is that virtualized systems take a significant performance penalty. This attack on virtualization generally is not relevant in the cloud world for a few reasons:

- The degraded performance of your cloud vendor's hardware is probably better than the optimal performance of your commodity server.
- Enterprise virtualization technologies such as Xen and VMware use paravirtualization as well as the hardware-assisted virtualization capabilities of a variety of CPU manufacturers to achieve near-native performance.

Cloud storage

Abstracting your hardware in the cloud is not simply about replacing servers with virtualization. It's also about replacing your physical storage systems.

Cloud storage enables you to "throw" data into the cloud and without worrying about how it is stored or backing it up. When you need it again, you simply reach into the cloud and grab it. You don't know how it is stored, where it is stored, or what has happened to all the pieces of hardware between the time you put it in the cloud and the time you retrieved it.

As with the other elements of cloud computing, there are a number of approaches to cloud storage on the market. In general, they involve breaking your data into small chunks and storing that data across multiple servers with fancy checksums so that the data can be retrieved

† Other approaches to cloud infrastructure exist, including physical hardware on-demand through companies such as AppNexus and NewClouds. In addition, providers such as GoGrid (summarized in Appendix B) offer hybrid solutions.

rapidly—no matter what has happened in the meantime to the storage devices that comprise the cloud.

I have seen a number of people as they get started with the cloud attempt to leverage cloud storage as if it were some kind of network storage device. Operationally, cloud storage and traditional network storage serve very different purposes. Cloud storage tends to be much slower with a higher degree of structure, which often renders it impractical for runtime storage for an application, regardless of whether that application is running in the cloud or somewhere else.

Cloud storage is not, generally speaking, appropriate for the operational needs of transactional cloud-based software. Later, we discuss in more detail the role of cloud storage in transaction application management. For now, think of cloud storage as a tape backup system in which you never have to manage any tapes.

> NOTE
>
> **Amazon recently introduced a new offering called Amazon CloudFront, which leverages Amazon S3 as a content distribution network. The idea behind Amazon CloudFront is to replicate your cloud content to the edges of the network. While Amazon S3 cloud storage may not be appropriate for the operational needs of many transactional web applications, CloudFront will likely prove to be a critical component to the fast, worldwide distribution of static content.**

Cloud Application Architectures

We could spend a lot of precious paper discussing Software as a Service or virtualization technologies (did you know that you can mix and match at least five kinds of virtualization?), but the focus of this book is how you write an application so that it can best take advantage of the cloud.

Grid Computing

Grid computing is the easiest application architecture to migrate into the cloud. A grid computing application is processor-intensive software that breaks up its processing into small chunks that can then be processed in isolation.

If you have used SETI@home, you have participated in grid computing. SETI (the Search for Extra-Terrestrial Intelligence) has radio telescopes that are constantly listening to activity in space. They collect volumes of data that subsequently need to be processed to search for a nonnatural signal that might represent attempts at communication by another civilization. It would take so long for one computer to process all of that data that we might as well wait until we can travel to the stars. But many computers using only their spare CPU cycles can tackle the problem extraordinarily quickly.

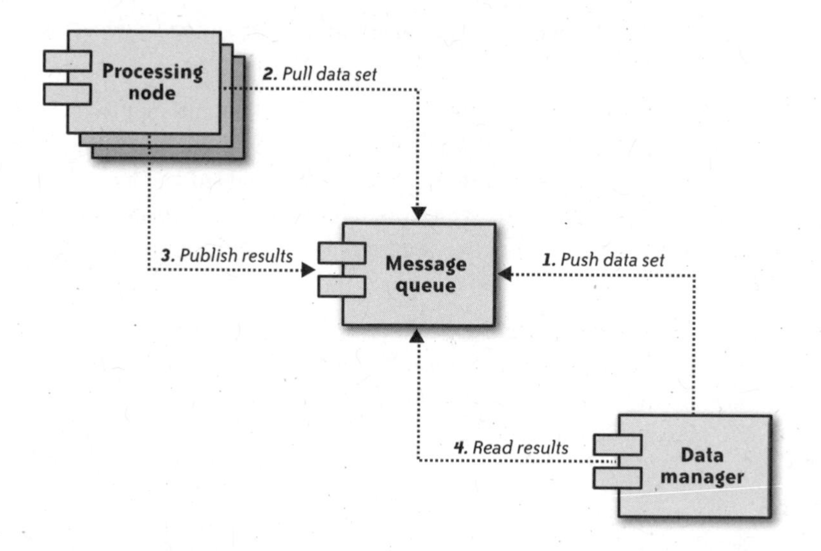

FIGURE 1-1. The grid application architecture separates the core application from its data processing nodes

These computers running SETI@home—perhaps including your desktop—form the grid. When they have extra cycles, they query the SETI servers for data sets. They process the data sets and submit the results back to SETI. Your results are double-checked against processing by other participants, and interesting results are further checked.‡

Back in 1999, SETI elected to use the spare cycles of regular consumers' desktop computers for its data processing. Commercial and government systems used to network a number of supercomputers together to perform the same calculations. More recently, *server farms* were created for grid computing tasks such as video rendering. Both supercomputers and server farms are very expensive, capital-intensive approaches to the problem of grid computing.

The cloud makes it cheap and easy to build a grid computing application. When you have data that needs to be processed, you simply bring up a server to process that data. Afterward, that server can either shut down or pull another data set to process.

Figure 1-1 illustrates the process flow of a grid computing application. First, a server or server cluster receives data that requires processing. It then submits that job to a message queue (1). Other servers—often called workers (or, in the case of SETI@home, other desktops)—watch the message queue (2) and wait for new data sets to appear. When a data set appears, the first computer to see it processes it and then sends the results back into the message queue (3). The two components can operate independently of each other, and one can even be running when no computer is running the other.

‡ For more information on SETI@home and the SETI project, pick up a copy of O'Reilly's *Beyond Contact* (*http://oreilly.com/catalog/9780596000370*).

Cloud computing comes to the rescue here because you do not need to own any servers when you have no data to process. You can then scale the number of servers to support the number of data sets that are coming into your application. In other words, instead of having idle computers process data as it comes in, you have servers turn themselves on as the rate of incoming data increases, and turn themselves off as the data rate decreases.

Because grid computing is currently limited to a small market (scientific, financial, and other large-scale data crunchers), this book doesn't focus on its particular needs. However, many of the principles in this book are still applicable.

Transactional Computing

Transactional computing makes up the bulk of business software and is the focus of this book. A *transaction system* is one in which one or more pieces of incoming data are processed together as a single transaction and establish relationships with other data already in the system. The core of a transactional system is generally a relational database that manages the relations among all of the data that make up the system.

Figure 1-2 shows the logical layout of a high-availability transactional system. Under this kind of architecture, an application server typically models the data stored in the database and presents it through a web-based user interface that enables a person to interact with the data. Most of the websites and web applications that you use every day are some form of transactional system. For high availability, all of these components may form a cluster, and the presentation/business logic tier can hide behind a load balancer.

Deploying a transactional system in the cloud is a little more complex and less obvious than deploying a grid system. Whereas nodes in a grid system are designed to be short-lived, nodes in a transactional system must be long-lived.

A key challenge for any system requiring long-lived nodes in a cloud infrastructure is the basic fact that the mean time between failures (MTBF) of a virtual server is necessarily less than that for the underlying hardware. An admittedly gross oversimplification of the problem shows that if you have two physical servers with a three-year MTBF, you will be less likely to experience an outage across the entire system than you would be with a single physical server running two virtual nodes. The number of physical nodes basically governs the MTBF, and since there are fewer physical nodes, there is a higher MTBF for any given node in your cloud-based transactional system.

The cloud, however, provides a number of avenues that not only help mitigate the lower failure rate of individual nodes, but also potentially increase the overall MTBF for your transactional system. In this book, we cover the tricks that will enable you to achieve levels of availability that otherwise might not be possible under your budget while still maintaining transactional integrity of your cloud applications.

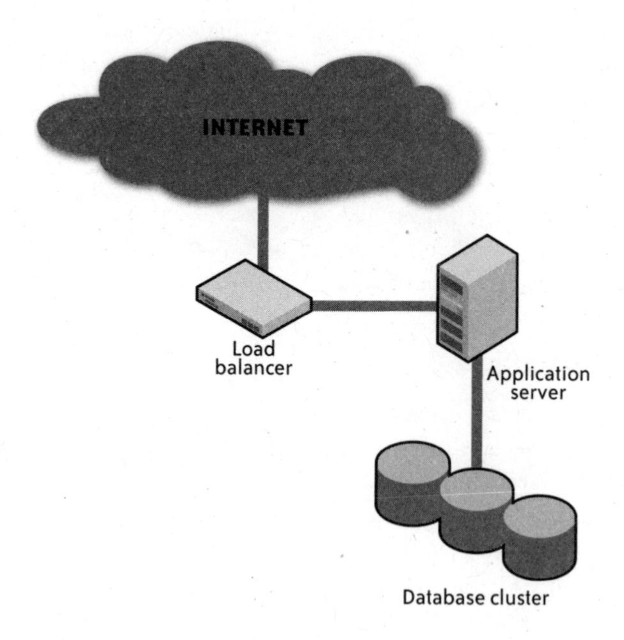

FIGURE 1-2. A transactional application separates an application into presentation, business logic, and data storage

The Value of Cloud Computing

How far can you take all of this?

If you can deploy all of your custom-built software systems on cloud hardware and leverage SaaS systems for your packaged software, you might be able to achieve an all-cloud IT infrastructure. Table 1-1 lists the components of the typical small- or medium-sized business.

TABLE 1-1. The old IT infrastructure versus the cloud

Traditional	Cloud
File server	Google Docs
MS Outlook, Apple Mail	Gmail, Yahoo!, MSN
SAP CRM/Oracle CRM/Siebel	SalesForce.com
Quicken/Oracle Financials	Intacct/NetSuite
Microsoft Office/Lotus Notes	Google Apps
Stellent	Valtira
Off-site backup	Amazon S3
Server, racks, and firewall	Amazon EC2, GoGrid, Mosso

The potential impact of the cloud is significant. For some organizations—particularly small- to medium-sized businesses—it makes it possible to never again purchase a server or own any software licenses. In other words, all of these worries diminish greatly or disappear altogether:

- Am I current on all my software licenses? SaaS systems and software with cloud-friendly licensing simply charge your credit card for what you use.
- When do I schedule my next software upgrade? SaaS vendors perform the upgrades for you; you rarely even know what version you are using.
- What do I do when a piece of hardware fails at 3 a.m.? Cloud infrastructure management tools are capable of automating even the most traumatic disaster recovery policies.
- How do I manage my technology assets? When you are in the cloud, you have fewer technology assets (computers, printers, etc.) to manage and track.
- What do I do with my old hardware? You don't own the hardware, so you don't have to dispose of it.
- How do I manage the depreciation of my IT assets? Your costs are based on usage and thus don't involve depreciable expenses.
- When can I afford to add capacity to my infrastructure? In the cloud, you can add capacity discretely as the business needs it.

SaaS vendors (whom I've included as part of cloud computing) can run all their services in a hardware cloud provided by another vendor, and therefore offer a robust cloud infrastructure to their customers without owning their own hardware. In fact, my own business runs that way.

Options for an IT Infrastructure

The cloud competes against two approaches to IT:

- Internal IT infrastructure and support
- Outsourcing to managed services

If you own the boxes, you have an internally managed IT infrastructure—even if they are sitting in a rack in someone else's data center. For you, the key potential benefit of cloud computing (certainly financially) is the lack of capital investment required to leverage it.

Internal IT infrastructure and support is one in which you own the boxes and pay people—whether staff or contract employees—to maintain those boxes. When a box fails, you incur that cost, and you have no replacement absent a cold spare that you own.

Managed services outsourcing has similar benefits to the cloud in that you pay a fixed fee for someone else to own your servers and make sure they stay up. If a server goes down, it is the managed services company who has to worry about replacing it immediately (or within whatever terms have been defined in your service-level agreement). They provide the expertise to make sure the servers are fixed with the proper operating system patches and manage the network infrastructure in which the servers operate.

Table 1-2 provides a comparison between internal IT, managed services, and cloud-based IT with respect to various facets of IT infrastructure development.

TABLE 1-2. A comparison of IT infrastructure options

	Internal IT	Managed services	The cloud
Capital investment	Significant	Moderate	Negligible
	How much cash do you have to cough up in order to set up your infrastructure or make changes to it? With internal IT, you have to pay for your hardware before you need it (financing is not important in this equation).[a] Under managed services, you are typically required to pay a moderate setup fee. In the cloud, you generally have no up-front costs and no commitment.		
Ongoing costs	Moderate	Significant	Based on usage
	Your ongoing costs for internal IT are based on the cost of staff and/or contractors to manage the infrastructure, as well as space at your hosting provider and/or real estate and utilities costs. You can see significant variances in the ongoing costs—especially with contract resources—as emergencies occur and other issues arise. Although managed services are often quite pricey, you generally know exactly what you are going to pay each month and it rarely varies. The cloud, on the other hand, can be either pricey or cheap, depending on your needs. Its key advantage is that you pay for exactly what you use and nothing more. Your staff costs are greater than with a managed services provider, but less than with internal IT.		
Provisioning time	Significant	Moderate	None
	How long does it take to add a new component into your infrastructure? Under both the internal IT and managed services models, you need to plan ahead of time, place an order, wait for the component to arrive, and then set it up in the data center. The wait is typically significantly shorter with a managed services provider, since they make purchases ahead of time in bulk. Under the cloud, however, you can have a new "server" operational within minutes of deciding you want it.		
Flexibility	Limited	Moderate	Flexible
	How easily can your infrastructure adapt to unexpected peaks in resource demands? For example, do you have a limit on disk space? What happens if you suddenly approach that limit? Internal IT has a very fixed capacity and can meet increased resource demands only through further capital investment. A managed services provider, on the other hand, usually can offer temporary capacity relief by uncapping your bandwidth, giving you short-term access to alternative storage options, and so on. The cloud, however, can be set up to automatically add capacity into your infrastructure as needed, and to let go of that capacity when it is no longer required.		

	Internal IT	Managed services	The cloud
Staff expertise requirements	Significant	Limited	Moderate
	How much expertise do you need in-house to support your environments? With internal IT, you obviously need staff or contractors who know the ins and outs of your infrastructure, from opening the boxes up and fiddling with the hardware to making sure the operating systems are up-to-date with the latest patches. The advantage here goes to the managed services infrastructure, which enables you to be largely ignorant of all things IT. Finally, the cloud may require a lot of skill or very little skill, depending on how you are using it. You can often find a cloud infrastructure manager (enStratus or RightScale, for example) to manage the environment, but you still must have the skills to set up your machine images.		
Reliability	Varies	High	Moderate to high
	How certain are you that your services will stay up 24/7? The ability to create a high-availability infrastructure with an internal IT staff is a function of the skill level of your staff and the amount of cash you invest in the infrastructure. A managed services provider is the safest, most proven alternative, but this option can lack the locational redundancy of the cloud. A cloud infrastructure, finally, has significant locational redundancies but lacks a proven track record of stability.		

[a] From a financial perspective, the difference between coughing up cash today and borrowing it from a bank is inconsequential. Either way, spending $40K costs you money. If you borrow it, you pay interest. If you take it out of your bank account, you lose the opportunity to do something else with it (cost of capital).

The one obvious fact that should jump out of this chart is that building an IT infrastructure from scratch no longer makes any sense. The only companies that should have an internal IT are organizations with a significant preexisting investment in internal IT or with regulatory requirements that prevent data storage in third-party environments.

Everyone else should be using a managed services provider or the cloud.

The Economics

Perhaps the biggest benefit of cloud computing over building out your own IT infrastructure has nothing to do with technology—it's financial. The "pay for what you use" model of cloud computing is significantly cheaper for a company than the "pay for everything up front" model of internal IT.

Capital costs

The primary financial problem with an internally based IT infrastructure is the *capital cost*. A capital cost is cash you pay for assets prior to their entering into operations. If you buy a server, that purchase is a capital cost because you pay for it all up front, and then you realize its benefits (in other words, you use it) over the course of 2–3 years.

Let's look at the example of a $5,000 computer that costs $2,000 to set up. The $5,000 is a capital cost and the $2,000 is a one-time expense. From an accounting perspective, the $5,000 cost is just a "funny money" transaction, in that $5,000 is moved from one asset account (your bank account) into another asset account (your fixed assets account). The $2,000, on the other hand, is a real expense that offsets your profits.

The server is what is called a *depreciable asset*. As it is used, the server is depreciated in accordance with how much it has been used. In other words, the server's value to the company is reduced each month it is in use until it is worth nothing and removed from service. Each reduction in value is considered an expense that offsets the company's profits.

Finance managers hate capital costs for a variety of reasons. In fact, they hate any expenses that are not tied directly to the current operation of the company. The core rationale for this dislike is that you are losing cash today for a benefit that you will receive slowly over time (technically, over the course of the depreciation of the server). Any business owner or executive wants to focus the organization's cash on things that benefit them today. This concern is most acute with the small- and medium-sized business that may not have an easy time walking into the bank and asking for a loan.

The key problem with this delayed realization of value is that money costs money. A company will often fund their operational costs through revenues and pay for capital expenses through loans. If you can grow the company faster than the cost of money, you win. If you cannot grow that rapidly or—worse—you cannot get access to credit, the capital expenses become a significant drain on the organization.

Cost comparison

Managed services infrastructures and the cloud are so attractive to companies because they largely eliminate capital investment and other up-front costs. The cloud has the added advantage of tying your costs to exactly what you are using, meaning that you can often connect IT costs to revenue instead of treating them as overhead.

Table 1-3 compares the costs of setting up an infrastructure to support a single "moderately high availability" transactional web application with a load balancer, two application servers, and two database servers. I took typical costs at the time of writing, October 2008.

TABLE 1-3. Comparing the cost of different IT infrastructures

	Internal IT	Managed services	The cloud
Capital investment	$40,000	$0	$0
Setup costs	$10,000	$5,000	$1,000
Monthly service fees	$0	$4,000	$2,400
Monthly staff costs	$3,200	$0	$1,000
Net cost over three years	$149,000	$129,000	$106,000

Table 1-3 makes the following assumptions:

- The use of fairly standard 1u server systems, such as a Dell 2950 and the high-end Amazon instances.
- The use of a hardware load balancer in the internal IT and managed services configuration and a software load balancer in the cloud.
- No significant data storage or bandwidth needs (different bandwidth or storage needs can have a significant impact on this calculation).
- The low end of the cost spectrum for each of the options (in particular, some managed services providers will charge up to three times the costs listed in the table for the same infrastructure).
- Net costs denominated in today's dollars (in other words, don't worry about inflation).
- A cost of capital of 10% (cost of capital is what you could have done with all of the up-front cash instead of sinking it into a server and setup fees—basically the money's interest rate plus opportunity costs).
- The use of third-party cloud management tools such as enStratus or RightScale, incorporated into the cloud costs.
- Staff costs representing a fraction of an individual (this isolated infrastructure does not demand a full-time employee under any model).

Perhaps the most controversial element of this analysis is what might appear to be an "apples versus oranges" comparison on the load balancer costs. The reality is that this architecture doesn't really require a hardware load balancer except for extremely high-volume websites. So you likely could get away with a software load balancer in all three options.

A software load balancer, however, is very problematic in both the internal IT and managed services infrastructures for a couple of reasons:

- A normal server is much more likely to fail than a hardware load balancer. Because it is much harder to replace a server in the internal IT and managed services scenarios, the loss of that software load balancer is simply unacceptable in those two scenarios, whereas it would go unnoticed in the cloud scenario.
- If you are investing in actual hardware, you may want a load balancer that will grow with your IT needs. A hardware load balancer is much more capable of doing that than a software load balancer. In the cloud, however, you can cheaply add dedicated software load balancers, so it becomes a nonissue.

In addition, some cloud providers (GoGrid, for example) include free hardware load balancing, which makes the entire software versus hardware discussion moot. Furthermore, Amazon is scheduled to offer its own load-balancing solution at some point in 2009. Nevertheless, if you don't buy into my rationale for comparing the hardware load balancers against the software load balancers, here is the comparison using all software load balancers: $134K for internal IT, $92K for managed services, and $106K for a cloud environment.

The bottom line

If we exclude sunk costs, the right managed services option and cloud computing are always financially more attractive than managing your own IT. Across all financial metrics—capital requirements, total cost of ownership, complexity of costs—internal IT is always the odd man out.

As your infrastructure becomes more complex, determining whether a managed services infrastructure, a mixed infrastructure, or a cloud infrastructure makes more economic sense becomes significantly more complex.

If you have an application that you know has to be available 24/7/365, and even 1 minute of downtime in a year is entirely unacceptable, you almost certainly want to opt for a managed services environment and not concern yourself too much with the cost differences (they may even favor the managed services provider in that scenario).

On the other hand, if you want to get high-availability on the cheap, and 99.995% is good enough, you can't beat the cloud.

URQUHART ON BARRIERS TO EXIT

In November 2008, James Urquhart and I engaged in a Twitter discussion[§] relating to the total cost of ownership of cloud computing (James is a market manager for the Data Center 3.0 strategy at Cisco Systems and member of the CNET blog network). What we realized is that I was looking at the problem from the perspective of starting with a clean slate; James was looking at the problem from the reality of massive existing investments in IT. What follows is a summary of our discussion that James has kindly put together for this book.

While it is easy to get enthusiastic about the economics of the cloud in "green-field" comparisons, most modern medium-to-large enterprises have made a significant investment in IT infrastructure that must be factored into the cost of moving to the cloud.

These organizations already own the racks, cooling, and power infrastructure to support new applications, and will not incur those capital costs anew. Therefore, the cost of installing and operating additional servers will be significantly less than in the examples.

In this case, these investments often tip the balance, and it becomes much cheaper to use existing infrastructure (though with some automation) to deliver relatively stable capacity loads. This existing investment in infrastructure therefore acts almost as a "barrier-to-exit" for such enterprises considering a move to the cloud.

§ *http://blog.jamesurquhart.com/2008/12/enterprise-barrier-to-exit-to-cloud.html*

Of course, there are certain classes of applications that even a large enterprise will find more cost effective to run in the cloud. These include:

- Applications with widely varying loads, for which peak capacity is hard to predict, and for which purchasing for peak load would be inefficient most of the time anyway.
- Applications with occasional or periodic load spikes, such as tax processing applications or retail systems hit hard prior to the holidays. The cloud can provide excess capacity in this case, through a technique called "cloudbursting."
- New applications of the type described in this book that would require additional data center space or infrastructure investment, such as new cooling or power systems.

It seems to me highly ironic—and perhaps somewhat unique—that certain aspects of the cloud computing market will be blazed not by organizations with multiple data centers and thousands upon thousands of servers, but by the small business that used to own a few servers in a server hotel somewhere that finally shut them down and turned to Amazon. How cool is that?

Cloud Infrastructure Models

We have talked about a number of the technologies that make up cloud computing and the general value proposition behind the cloud. Before we move into building systems in the cloud, we should take a moment to understand a variety of cloud infrastructure models. I will spend the most time on the one most people will be working with, Amazon Web Services. But I also touch on a few of the other options.

It would be easy to contrast these services if there were fine dividing lines among them, but instead, they represent a continuum from managed services through something people call Infrastructure as a Service (IaaS) to Platform as a Service (PaaS).

Platform As a Service Vendor

PaaS environments provide you with an infrastructure as well as complete operational and development environments for the deployment of your applications. You program using the vendor's specific application development platform and let the vendor worry about all deployment details.

The most commonly used example of pure PaaS is Google App Engine. To leverage Google App Engine, you write your applications in Python against Google's development frameworks with tools for using the Google filesystem and data repositories. This approach works well for applications that must be deployed rapidly and don't have significant integration requirements.

The downside to the PaaS approach is vendor lock-in. With Google, for example, you must write your applications in the Python programming language to Google-specific APIs.

Python is a wonderful programming language—in fact, my favorite—but it isn't a core competency of most development teams. Even if you have the Python skills on staff, you still must contend with the fact that your Google App Engine application may only ever work well inside Google's infrastructure.

Infrastructure As a Service

The focus of this book is the idea of IaaS. I spend a lot of time in this book using examples from the major player in this environment, Amazon Web Services. A number of significant AWS competitors exist who have different takes on the IaaS problem. These different approaches have key value propositions for different kinds of cloud customers.

AWS is based on pure virtualization. Amazon owns all the hardware and controls the network infrastructure, and you own everything from the guest operating system up. You request virtual instances on-demand and let them go when you are done. Amazon sees one of its key benefits is a commitment to not overcommitting resources to virtualization.

AppNexus represents a different approach to this problem. As with AWS, AppNexus enables you to gain access to servers on demand. AppNexus, however, provides dedicated servers with virtualization on top. You have the confidence in knowing that your applications are not fighting with anyone else for resources and that you can meet any requirements that demand full control over all physical server resources.

Hybrid computing takes advantage of both worlds, offering virtualization where appropriate and dedicated hardware where appropriate. In addition, most hybrid vendors such as Rackspace and GoGrid base their model on the idea that people still want a traditional data center—they just want it in the cloud.

As we examine later in this book, there are a number of reasons why a purely virtualized solution might not work for you:

- Regulatory requirements that demand certain functions operate on dedicated hardware
- Performance requirements—particularly in the area of I/O—that will not support portions of your application
- Integration points with legacy systems that may lack any kind of web integration strategy

A cloud approach tied more closely to physical hardware may meet your needs in such cases.

Private Clouds

I am not a great fan of the term *private clouds*, but it is something you will often hear in reference to on-demand virtualized environments in internally managed data centers. In a private cloud, an organization sets up a virtualization environment on its own servers, either in its own data centers or in those of a managed services provider. This structure is useful for companies that either have significant existing IT investments or feel they absolutely must have total control over every aspect of their infrastructure.

The key advantage of private clouds is control. You retain full control over your infrastructure, but you also gain all of the advantages of virtualization. The reason I am not a fan of the term "private cloud" is simply that, based on the criteria I defined earlier in this chapter, I don't see a private cloud as a true cloud service. In particular, it lacks the freedom from capital investment and the virtually unlimited flexibility of cloud computing. As James Urquhart noted in his "Urquhart on Barriers to Exit" on page 16, I also believe that private clouds may become an excuse for not moving into the cloud, and could thus put the long-term competitiveness of an organization at risk.

All of the Above

And then there is Microsoft Azure. Microsoft Azure represents all aspects of cloud computing, from private clouds up to PaaS. You write your applications using Microsoft technologies and can deploy them initially in a private cloud and later migrate them to a public cloud. Like Google App Engine, you write applications to a proprietary application development framework. In the case of Azure, however, the framework is based on the more ubiquitous .NET platform and is thus more easily portable across Microsoft environments.

An Overview of Amazon Web Services

My goal in this book is to stick to general principles you can apply in any cloud environment. In reality, however, most of you are likely implementing in the AWS environment. Ignoring that fact is just plain foolish; therefore, I will be using that AWS environment for the examples used throughout this book.

AWS is Amazon's umbrella description of all of their web-based technology services. It encompasses a wide variety of services, all of which fall into the concept of cloud computing (well, to be honest, I have no clue how you categorize Amazon Mechanical Turk). For the purposes of this book, we will leverage the technologies that fit into their Infrastructure Services:

- Amazon Elastic Cloud Compute (Amazon EC2)
- Amazon Simple Storage Service (Amazon S3)
- Amazon Simple Queue Service (Amazon SQS)
- Amazon CloudFront
- Amazon SimpleDB

Two of these technologies—Amazon EC2 and Amazon S3—are particularly interesting in the context of transactional systems.

As I mentioned earlier, message queues are critical in grid computing and are also useful in many kinds of transactional systems. They are not, however, typical across web applications, so Amazon SQS will not be a focus in this book.

Given that the heart of a transactional system is a database, you might think Amazon SimpleDB would be a critical piece for a transactional application in the Amazon cloud. In reality, however, Amazon SimpleDB is—as its name implies—simple. Therefore, it's not well suited to large-scale web applications. Furthermore, it is a proprietary database system, so an application too tightly coupled to Amazon SimpleDB is stuck in the Amazon cloud.

Amazon Elastic Cloud Compute (EC2)

Amazon EC2 is the heart of the Amazon cloud. It provides a web services API for provisioning, managing, and deprovisioning virtual servers inside the Amazon cloud. In other words, any application anywhere on the Internet can launch a virtual server in the Amazon cloud with a single web services call.

At the time of this writing, Amazon's EC2 U.S. footprint spans three data centers on the East Coast of the U.S. and two in Western Europe. You can sign up separately for an Amazon European data center account, but you cannot mix and match U.S. and European environments. The servers in these environments run a highly customized version of the Open Source Xen hypervisor using paravirtualization. This Xen environment enables the dynamic provisioning and deprovisioning of servers, as well as the capabilities necessary to provide isolated computing environment for guest servers.

When you want to start up a virtual server in the Amazon environment, you launch a new node based on a predefined Amazon machine image (AMI). The AMI includes your operating system and any other prebuilt software. Most people start with a standard AMI based on their favorite operating system, customize it, create a new image, and then launch their servers based on their custom images.

By itself, EC2 has two kinds of storage:

- Ephemeral storage tied to the node that expires with the node
- Block storage that acts like a SAN and persists across time

Many competitors to Amazon also provide persistent internal storage for nodes to make them operate more like a traditional data center.

In addition, servers in EC2—like any other server on the Internet—can access Amazon S3 for cloud-based persistent storage. EC2 servers in particular see both cost savings and greater efficiencies in accessing S3.

To secure your network within the cloud, you can control virtual firewall rules that define how traffic can be filtered to your virtual nodes. You define routing rules by creating security groups and associating the rules with those groups. For example, you might create a DMZ group that allows port 80 and port 443 traffic from the public Internet into its servers, but allows no other incoming traffic.

Amazon Simple Storage Service (S3)

Amazon S3 is cloud-based data storage accessible in real time via a web services API from anywhere on the Internet. Using this API, you can store any number of objects—ranging in size from 1 byte to 5 GB—in a fairly flat namespace.

It is very important *not* to think of Amazon S3 as a filesystem. I have seen too many people get in trouble when they expect it to act that way. First of all, it has a two-level namespace. At the first level, you have buckets. You can think of these buckets as directories, if you like, as they store the data you put in S3. Unlike traditional directories, however, you cannot organize them hierarchically—you cannot put buckets in buckets. Perhaps more significant is the fact that the bucket namespace is shared across all Amazon customers. You need to take special care in designing bucket names that will not clash with other buckets. In other words, you won't be creating a bucket called "Documents".

Another important thing to keep in mind is that Amazon S3 is relatively slow. Actually, it is very fast for an Internet-deployed service, but if you are expecting it to respond like a local disk or a SAN, you will be very disappointed. Therefore, it is not feasible to use Amazon S3 as an operational storage medium.

Finally, access to S3 is via web services, not a filesystem or WebDAV. As a result, applications must be written specifically to store data in Amazon S3. Perhaps more to the point, you can't simply *rsync* a directory with S3 without specially crafted tools that use the Amazon API and skirt the S3 limitations.

I have spent enough text describing what Amazon S3 is not—so what is it?

Amazon S3 enables you to place persistent data into the cloud and retrieve it at a later date with a near certainty that it will be there in one consistent piece when you get it back. Its key benefit is that you can simply continue to shove data into Amazon S3 and never worry about running out of storage space. In short, for most users, S3 serves as a short-term or long-term backup facility.

CLEVERSAFE STORAGE

Cloud storage systems have unique challenges that legacy storage technologies cannot address. Storage technologies based on RAID and replication are not well suited for cloud infrastructures because they don't scale easily to the exabyte level. Legacy storage technologies rely on redundant copies to increase reliability, resulting in systems that are not easily manageable, chew up bandwidth, and are not cost effective.

Cleversafe's unique cloud storage platform—based on company technology trademarked under the name Dispersed Storage—divides data into slices and stores them in different geographic locations on hardware appliances. The algorithms used to divide the data are comparable to the concept of parity—but with much more sophistication—because they allow the total data to be reconstituted from a subset. For instance, you may store the data in 12 locations, any 8 of which are enough to restore it completely. This technology, known as information dispersal, achieves geographic redundancy and high availability without expensive replication of the data.

In April 2008, Cleversafe embodied its dispersal technology in hardware appliances that provide a front-end to the user using standard protocols such as REST APIs or iSCSI. The appliances take on the task of splitting and routing the data to storage sites, and merely increase the original file size by 1.3 to 1.6 times, versus 3 times in a replicated system.

Companies are using Cleversafe's Dispersed Storage appliances to build public and private cloud storage as a backend infrastructure to Software as a Service. Dispersed Storage easily fulfills the characteristics of a cloud infrastructure since it provides storage on demand and accessibility anywhere.

Dispersal also achieves higher levels of security within the cloud without necessarily needing encryption, because each slice contains too little information to be useful. This unique architecture helps people satisfy their concern over their data being outside of their immediate control, which often becomes a barrier to storage decisions. While a lost backup tape contains a full copy of data, access to a single appliance using Dispersed Storage results in no data breach.

Additionally, Dispersed Storage is massively scalable and designed to handle petabytes of data. By adding servers into the storage cloud with automated storage discovery, the total storage of the system can easily grow, and performance can be scaled by simply adding additional appliances. Virtualization tools enable easy deployment and on-demand provisioning. All of these capabilities streamline efforts for storage administrators.

Dispersed Storage is also designed to store and distribute large objects, the cornerstone of our media-intensive society that has become dependent on videos and images in every aspect of life. Dispersal is inherently designed for content distribution by naturally incorporating load balancing through the multitude of access choices for selecting the slices used to reconstruct the original file. This means companies do not have to deal with or pay for implementing a separate content delivery network for their stored data.

Dispersed Storage offers a novel and needed approach to cloud storage, and will be significant as cloud storage matures and displaces traditional storage methods.

Amazon Simple Queue Service (SQS)

Amazon SQS is a cornerstone to any Amazon-based grid computing effort. As with any message queue service, it accepts messages and passes them on to servers subscribing to the message queue.

A messaging system typically enables multiple computers to exchange information in complete ignorance of each other. The sender simply submits a short message (up to 8KB in Amazon SQS) into the queue and continues about its business. The recipient retrieves the message from the queue and acts upon the contents of the message.

A message, for example, can be, "Process data set 123.csv in S3 bucket s3://fancy-bucket and submit the results to message queue Y." One advantage of a message queue system is that the sender does not need to identify a recipient or perform any error handling to deal with communication failures. The recipient does not even need to be active at the time the message is sent.

The Amazon SQS system fits well into a cloud computing environment due to its simplicity. Most systems requiring a message queue need only a simple API to submit a message, retrieve it, and trust the integrity of the message within the queue. It can be a tedious task to develop and maintain something this simple, but it is also way too complex and expensive to use many of the commercial message queue packages.

Amazon CloudFront

Amazon CloudFront, a cloud-based content distribution network (CDN), is a new offering from Amazon Web Services at the time of this writing. It enables you to place your online content at the edges of the network, meaning that content is delivered from a location close to the user requesting it. In other words, a site visitor from Los Angeles can grab the same content from an Amazon server in Los Angeles that a visitor from New York is getting from a server in New York. You place the content in S3 and it gets moved to the edge points of the Amazon network for rapid delivery to content consumers.

Amazon SimpleDB

Amazon SimpleDB is an odd combination of structured data storage with higher reliability than your typical MySQL or Oracle instance, and very baseline relational storage needs. It is very powerful for people concerned more with the availability of relational data and less so with the complexity of their relational model or transaction management. In my experience, this audience is a very small subset of transactional applications—though it could be particularly useful in heavy read environments, such as web content management systems.

The advantages of Amazon SimpleDB include:

- No need for a database administrator (DBA)
- A very simple web services API for querying the data
- Availability of a clustered database management system (DBMS)
- Very scalable in terms of data storage capabilities

If you need the power of a relational database, Amazon SimpleDB is not an appropriate tool. On the other hand, if your idea of an application database is *bdb,* Amazon SimpleDB will be the perfect tool for you.

CHAPTER TWO

Amazon Cloud Computing

AS I MENTIONED IN THE PREVIOUS CHAPTER, this book is a far-ranging, general guide for developers and systems administrators who are building transactional web applications in any cloud. As I write this book, however, the term "cloud infrastructure" is largely synonymous with Amazon EC2 and Amazon S3 for a majority of people working in the cloud. This reality combined with my use of Amazon cloud examples demands an overview of cloud computing specifically in the Amazon cloud.

Amazon S3

Amazon Simple Storage Service (S3) is cloud-based persistent storage. It operates independently from other Amazon services. In fact, applications you write for hosting on your own servers can leverage Amazon S3 without any need to otherwise "be in the cloud."

When Amazon refers to S3 as "simple storage," they are referring to the feature set—not its ease of use. Amazon S3 enables you to simply put data in the cloud and pull it back out. You do not need to know anything about how it is stored or where it is actually stored.

You are making a terrible mistake if you think of Amazon S3 as a remote filesystem. Amazon S3 is, in many ways, much more primitive than a filesystem. In fact, you don't really store "files"—you store objects. Furthermore, you store objects in buckets, not directories. Although these distinctions may appear to be semantic, they include a number of important differences:

- Objects stored in S3 can be no larger than 5 GB.
- Buckets exist in a flat namespace shared among all Amazon S3 users. You cannot create "sub-buckets," and you must be careful of namespace clashes.
- You can make your buckets and objects available to the general public for viewing.
- Without third-party tools, you cannot "mount" S3 storage. In fact, I am not fond of the use of third-party tools to mount S3, because S3 is so conceptually different from a filesystem that I believe it is bad form to treat it as such.

Access to S3

Before accessing S3, you need to sign up for an Amazon Web Services account. You can ask for default storage in either the United States or Europe. Where you store your data is not simply a function of where you live. As we discuss later in this book, regulatory and privacy concerns will impact the decision of where you want to store your cloud data. For this chapter, I suggest you just use the storage closest to where your access to S3 will originate.

Web Services

Amazon makes S3 available through both a SOAP API and a REST API. Although developers tend to be more familiar with creating web services via SOAP, REST is the preferred mechanism for accessing S3 due to difficulties processing large binary objects in the SOAP API. Specifically, SOAP limits the object size you can manage in S3 and limits any processing (such as a transfer status bar) you might want to perform on the data streams as they travel to and from S3.

The Amazon Web Services APIs support the ability to:

- Find buckets and objects
- Discover their metadata
- Create new buckets
- Upload new objects
- Delete existing buckets and objects

When manipulating your buckets, you can optionally specify the location in which the bucket's contents should be stored.

Unless you need truly fine-grained control over interaction with S3, I recommend using an API wrapper for your language of choice that abstracts out the S3 REST API. My teams use Jets3t when doing Java development.

For the purposes of getting started with Amazon S3, however, you will definitely want to download the *s3cmd* command-line client for Amazon S3 (*http://s3tools.logix.cz/s3cmd*). It provides a command-line wrapper around the S3 access web services. This tool also happens to be written in Python, which means you can read the source to see an excellent example of writing a Python application for S3.

BitTorrent

Amazon also provides BitTorrent access into Amazon S3. BitTorrent is a peer-to-peer (P2P) filesharing protocol. Because BitTorrent is a standard protocol for sharing large binary assets, a number of clients and applications exist on the market to consume and publish data via BitTorrent. If your application can leverage this built-in infrastructure, it may make sense to take advantage of the Amazon S3 BitTorrent support. In general, however, transactional web applications won't use BitTorrent to interact with S3.

S3 in Action

To illustrate S3 in action, we will use the *s3cmd* utility to transfer files in and out of S3. The commands supported by this tool are mirrors of the underlying web services APIs. Once you download the utility, you will need to configure it with your S3 access key and S3 secret key. Whether you are using this tool or another tool, you will always need these keys to access your private buckets in S3.

The first thing you must do with Amazon S3 is create a bucket in which you can store objects:

```
s3cmd mb s3://BUCKET
```

This command creates a bucket with the name you specify. As I noted earlier in the chapter, the namespace for your bucket is shared across all Amazon customers. Unless you are the first person to read this book, it is very unlikely that the command just shown will succeed unless you replace the name *BUCKET* with something likely to be unique to you. Many users prefix their buckets with something unique to them, such as their domain name. Keep in mind, however, that whatever standard you pick, nothing stops other users from stepping over your naming convention.

BUCKET NAMING

I have already cautioned you on a number of occasions regarding the flat bucket namespace. You are probably wondering how you can work under such naming constraints.

First, keep in mind the following naming rules:

- Names may contain only lowercase letters, numbers, periods, underscores, and dashes, and they must start with a number or a letter.
- A name cannot be in the style of an IP address (in other words, not 10.0.0.1).
- A name must be at least 3 characters long and be no more than 255 characters long.

You will want to name your buckets in such a way that you will create your own virtual bucket namespace. For example, `com.imaginary.`*mybucket* is likely to be available to me because I own the domain *imaginary.com*. It is important to note, however, that nothing guarantees that no one else will use your domain (or whatever other naming convention you use) as a prefix for their bucket names. Your applications therefore need to be intelligent when creating new buckets and should not be fixed to a particular naming scheme.

When naming a bucket, Amazon suggests the following naming guidelines to enhance your ability to create valid URLs for S3 objects:

- You should not name buckets using underscores (even though it is allowed).
- You should ideally limit bucket names to 63 characters.
- You should not end a bucket name with a dash, nor should you have a name that has a dash followed by a period.

Once you have created a bucket, it's time to stick things in it:

```
s3cmd put LOCAL_FILE s3://BUCKET/S3FILE
```

For example:

```
s3cmd put home_movie.mp4 s3://com.imaginary.movies/home_movie.mp4
```

The *s3cmd* utility limits your files to 5 GB in size because of the S3 limit mentioned earlier. Later in the book, we will discuss appropriate strategies for getting around this limitation and providing a higher level of security and integrity management around your S3 objects.

You can then get the object out of the cloud:

```
s3cmd get s3://BUCKET/S3FILE LOCAL_FILE
```

For example:

```
s3cmd get s3://com.imaginary.movies/home_movie.mp4 home_movies3.mp4
```

You should now have your home movie back on your desktop.

The following are other commands you can leverage:

- List all of your buckets: `s3cmd ls`
- List the contents of a specific bucket: `s3cmd ls s3://`*BUCKET*
- Delete an object from a bucket: `s3cmd del s3://`*BUCKET/S3FILE*
- Delete a bucket: `s3cmd rb s3://`*BUCKET*

You can delete a bucket only if it is empty. You must first delete the contents one by one and then delete the bucket. The *s3cmd* soon will have a *--recursive* option to the *del* command, but you should be aware that it is simply listing the contents of the target bucket and deleting

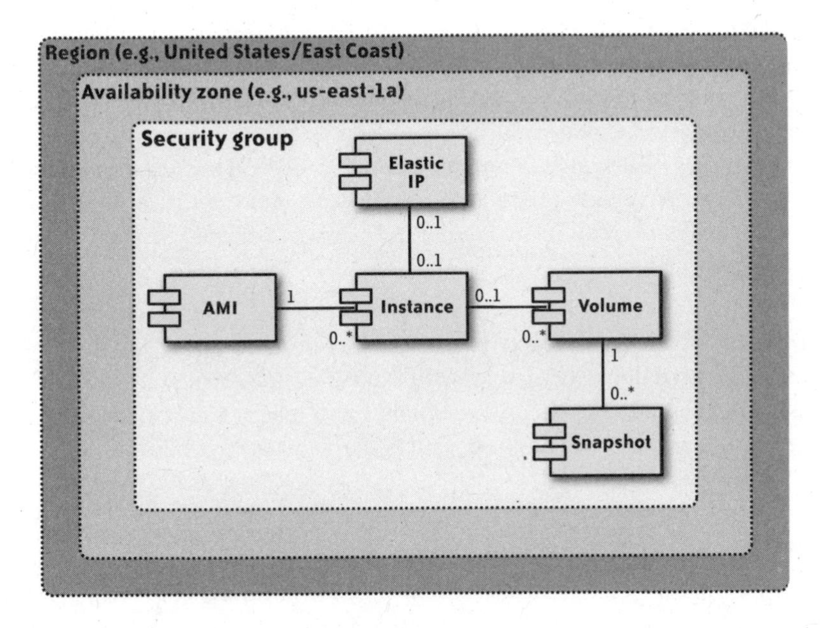

FIGURE 2-1. An overview of the components that support Amazon EC2

them one by one. If you are using the web services API, you will need to write your own recursive delete when deleting a bucket that contains objects.

Amazon EC2

When most people think of the Amazon cloud, they are thinking about Amazon EC2. EC2 represents your virtual network with all of the virtual servers running inside that network. It does not, however, stand alone. When you use EC2, you will be using S3 to store your machine images (more on that in a minute) and potentially for other storage needs. If you skipped the previous section, thinking you did not need to know about S3, go back and read it!

EC2 Concepts

EC2 is quite a bit more complex than S3. Figure 2-1 shows all of the concepts that make up Amazon EC2 and how they relate to each other.

The main concepts are:

Instance
: An EC2 instance is a virtual server running your choice of guest operating system based on the machine image from which the instance was cloned.

Amazon Machine Image (AMI)

A pristine copy of your server that you can use to launch any number of instances. If you are familiar with the concept of ghosting, the machine image represents your ghost image from which you can build any number of servers. Minimally, a machine image will have the core operating system plus common preinstalled tools. Depending on your deployment strategy, it might have your prebuilt web application. Amazon has prebuilt AMIs to get you started. In addition, there are many third-party AMIs, and you can build your own.

Elastic IP address

This is simply a static IP address that is assigned to you. (The term "elastic" may be confusing: rest assured that this term refers to a static address, not a dynamic one.) By default, each Amazon instance comes with a dynamically assigned IP address that may be reassigned to another user when your instance terminates. Elastic IP addresses are reserved to you and thus useful for instances that must always be accessible by the same static IP address.

Region

A group of availability zones that form a single geographic cluster. At the time of this book's publication, Amazon's service level agreement (SLA) for EC2 guarantees 99.95% availability of at least two availability zones within a region over the course of a 12-month period.

Availability zone

Roughly analogous to a data center. Two availability zones are guaranteed not to share any common points of failure. Amazon currently has three zones in the U.S., all on the East Coast. It also has two zones in Western Europe. You may optionally define the availability zone into which you launch your instances to create a level of locational redundancy for your applications.

Security group

Very roughly analogous to a network segment governed by a firewall. You launch your instances into security groups and, in turn, the security groups define what can talk to your new instances.

Block storage volume

Conceptually similar to a SAN, it provides block-level storage that you can mount from your EC2 instances. You can then format the volume as you like, or write raw to the volume. You can even leverage multiple volumes together as a virtual RAID.

Snapshot

You may take "snapshots" whenever you like of your block volumes for backup or replication purposes. These snapshots are stored in Amazon S3, where they can be used to create new volumes.

MACHINE IMAGES VERSUS INSTANCES

When getting started with Amazon EC2, you should not be discouraged if the distinction between machine images (AMIs) and instances is not entirely clear. It is very important, however, to understand the distinction because your machine image management strategy will be critical to your long-term success in the cloud.

A machine image is the prototype from which your virtual servers are created. Perhaps the best way to conceptualize a machine image is to think of it as a copy of your server's hard drive that gets cloned onto a new server just before it is launched. That "new server" is your instance. Each time you launch an instance, EC2 copies the machine image onto a new instance's virtual hard drive and then boots up the instance. You can create any number of running instances from a single machine image.

To use a developer analogy, if you think of your machine image as a class in a programming language, the EC2 instance is an object instance.

EC2 Access

Like Amazon S3, the primary means of accessing Amazon EC2 is through a web services API. Amazon provides a number of interactive tools on top of their web services API, including:

- The Amazon Web Services Console (*http://console.aws.amazon.com*; see Figure 2-2)
- The ElasticFox Firefox plug-in
- The Amazon Command Line tools

Once your use of Amazon matures, you will likely use robust infrastructure management tools such as enStratus and RightScale. As you are getting started, however, you will probably use a combination of the tools Amazon provides. The examples throughout the rest of this chapter focus on the command-line tools available at *http://developer.amazonwebservices.com/connect/entry.jspa?externalID=351*. Once you get the command-line tools down, everything else is simple. Regardless of the tools you ultimately use, however, you will always have a need for the command line.

Instance Setup

The simplest thing you can do in EC2 is launch an instance. This action dynamically sets up a server and makes it available for your use.

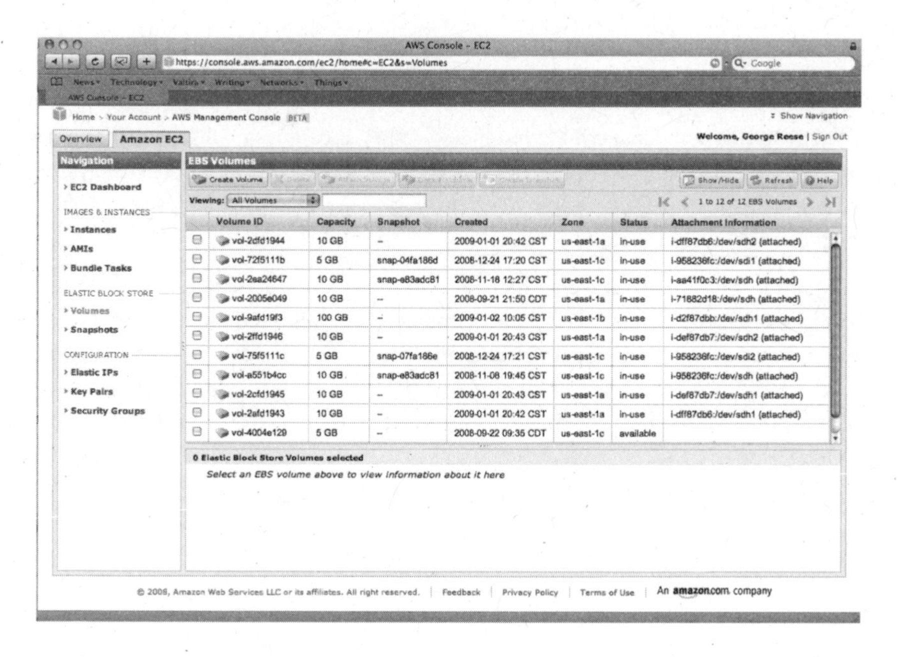

FIGURE 2-2. The Amazon Web Services Console

You launch an instance from an (AMI) stored in S3. Amazon provides a number of prebuilt machine images, and you can buy access to commercial ones. Once you have developed a level of comfort with EC2, you can even create and register your own machine images.

WARNING

Because your machine images are stored in Amazon S3, you cannot launch a new EC2 instance when Amazon S3 is unavailable.

Through the command-line utilities, you can see what machine images are readily available to you:

```
ec2-describe-images -o amazon
```

The `-o amazon` option tells the command to look for all machine images owned by Amazon. When you are getting started, you will likely leverage one of these standard images or a standard image from another vendor. Once you have customized those images, you will build your own images and register them. To see what images you have registered, enter:

```
ec2-describe-images
```

The output looks something like this:

```
IMAGE    ami-225fba4b    ec2-public-images/fedora-core4-apache-mysql-
v1.07.manifest.xml    amazon    available    public        i386    machine
```

The second element in the output is the image ID. You need this value to refer to the image in all other contexts, whether you are using the command-line tools or the web services API.

The third element is the S3 object that represents the machine image's manifest file. The manifest file describes the machine image to EC2. In this example, *ec2-public-images* is the bucket in which the image is stored, and *fedora-core4-apache-mysql-v1.07.manifest.xml* is the object representing the image's manifest file.

The fourth element is the owner of the machine image—in this case, Amazon. The fifth element describes the state of the image. The sixth element identifies whether the image is publicly available or private.

The seventh element, `i386`, identifies the target architecture for the image. Amazon currently supports two architectures: 32-bit Intel (i386) and 64-bit Intel (x86_64).

The final element identifies what kind of image this image represents: machine, ramdisk, or kernel. For now, we will concern ourselves only with machine images.

SELECTING AN AMI

The selection of a starter AMI in EC2 used to be simple. Today, however, you have a wide variety of Amazon-built and third-party AMIs from which to choose. It's often hard to tell which AMIs are reputable, which are junk, and which might include Trojans or backdoors. Fortunately, Amazon does provide a catalog with reviews and ratings at *http://aws.amazon.com/amis*. The starting point is often your favorite flavor of Linux (though other kinds of Unix are currently available, and even Windows). Identify someone who has put together a solid, minimalist version of your distribution targeted at your desired architecture and leverage that AMI as your base. I am personally fond of the Ubuntu builds created by Eric Hammond.

Whatever you choose, you should start out as basic as possible and gradually install only the utilities and services that are absolutely necessary to support your web application. Before going nuts with installations, harden your server using a tool such as Bastille (*http://www.bastille-unix.org*) and then revisit the hardening once you are done installing custom applications. For more details on securing your Amazon instances, see Chapter 5.

Once you have picked out the right machine image, it is time to launch a virtual server instance based on that machine image:

```
ec2-run-instances AMI_ID
```

For example:

```
$ ec2-run-instances ami-1fd73376
```

You will receive either an error message or information about the *reservation* for your new instance:

```
RESERVATION     r-3d01de54     1234567890123     default
INSTANCE     i-b1a21bd8     ami-1fd73376              pending        0
    m1.small     2008-10-22T16:10:38+0000     us-east-1a     aki-a72cf9ce
    ari-a52cf9cc
```

The distinction between a reservation and instance is not so obvious in this trivial example, in which I have started a single instance. When you attempt to run instances, you can specify a minimum and maximum number of instances to launch—not just a single instance. The command returns with a reservation that describes how many instances could actually be launched with that one request.

For example, you may be launching new application servers to help distribute website load as a heavy part of the day is approaching. If you want to launch 10 instances to support the increased demand, you may encounter the possibility that the resources for all 10 instances may not be available at exactly that instant in any single availability zone. If you requested 10 when not all 10 were available, the request would fail. You can, however, request a minimum of 5 with a maximum of 10. As long as enough resources for at least 5 instances are available, the command will succeed. The reservation thus describes the successfully launched instances.

Some important things to note from the instance description include:

`i-b1a21bd8`
: The instance ID. You use it with other commands and the web services API to refer to the instance.

`ami-1fd73376`
: The machine image ID of the AMI you used to launch the instance.

`pending`
: The current state of the instance. Other possible values are `running`, `shutting down`, and `terminated`.

`m1.small`
: Which EC2 instance type you are running. The instance types specify how much RAM, disk space, and CPU your instance has allocated to it. The `m1.small` type is the very baseline instance.

`us-east-1a`
: The availability zone into which the instance was launched. Because we did not specify an availability zone at the command line, EC2 picked the best availability zone to launch into based on current usage. I will dive deeper into availability zones later in this chapter.

At this point, you wait. It may take up to 10 minutes for your instance to become available. While you wait, Amazon is allocating the resources for the instance and then booting it up. You can check on the status using the *ec2-describe-instances* command:

```
ec2-describe-instances [INSTANCE_ID]
```

For example:

```
$ ec2-describe-instances i-b1a21bd8
RESERVATION    r-3d01de54    1234567890123    default
INSTANCE    i-b1a21bd8    ami-1fd73376    ec2-75-101-201-11.compute-1.amazonaws.com
    domU-12-31-38-00-9D-44.compute-1.internal    running        0
    m1.small    2008-08-11T14:39:09+0000    us-east-1c    aki-a72cf9ce
    ari-a52cf9cc
```

This output shows that the instance is now in a "running" state. In other words, you can access it, but there is one little problem: you don't have a user account on the machine.

Before I tell you how to get a shell account on the machine, you should first understand what other information you have about the instance now that it is running.

The output of the *ec2-describe-instances* command is almost identical to the *ec2-run-instances* command, except that it now provides you with the public IP address and private IP address of the instance because the instance is now in a "running" state. When an instance is launched, Amazon will dynamically assign one public and one private IP address to that instance. You should use the public IP address to access the instance from outside the cloud and the private address to access it from inside the cloud.

Access to an Instance

As I mentioned already, you won't be able to access a newly launched EC2 instance until you have an account on it. So let's terminate it and launch a different instance to which you will have access:

```
$ ec2-terminate-instances i-b1a21bd8
```

Since you are using a prebuilt machine instance, you obviously will not have any user accounts on that image when it comes up. Furthermore, it would be a huge security hole for Amazon (or anyone else) to create default accounts with common passwords like "scott"/"tiger".

The trick is to generate an SSH keypair.The private key will sit on your local hard drive, whereas the public key is passed to the instance when you launch it. EC2 will configure the instance so that the "root" account is accessible to any user with your private key.

Before you launch the new instance, you need to generate a keypair:

```
ec2-add-keypair KEYPAIR_NAME
```

For example:

```
$ ec2-add-keypair mykeypair
KEYPAIR mykeypair  1f:51:ae:28:bf:89:e9:d8:1f:25:5d:37:2d:7d:b8:ca:9f:f5:f1:6f
-----BEGIN RSA PRIVATE KEY-----
MIIEoQIBAAKCAQBuLFg5ujHrtm1jnutSuoO8Xe56LlT+HM8v/xkaa39EstM3/aFxTHgElQiJLChp
HungXQ29VTc8rc1bW0lkdi23OH5eqkMHGhvEwqaOHWASUMll4o3o/IX+0f2UcPoKCOVUR+jx71Sg
5AU52EQfanIn3ZQ8lFW7Edp5a3q4DhjGlUKToHVbicL5E+g45zfB95wIyywWZfeW/UUF3LpGZyq/
ebIUlq1qTbHkLbCC2r7RTn8vpQWp47BGVYGtGSBMpTRP5hnbzzuqj3itkiLHjU39S2sJCJ0TrJx5
i8BygR4s3mHKBj8l+ePQxG1kGbF6R4yg6sECmXn17MRQVXODNHZbAgMBAAECggEAY1tsiUsIwDl5
```

```
91CXirkYGuVfLyLflXenxfI5OmDFms/mumTqloHO7trOoriHDR5K7wMcY/YY5YkcXNo7mvUVD1pM
ZNUJs7rw9gZRTrf7LylaJ58kOcyajw8TsC4e4LPbFaHwS1d6K8rXh64o6WgW4SrsB6ICmr1kGQI7
3wcfgt5ecIu4TZfOOE9IHjn+2eRlsrjBdeORi7KiUNC/pAG23I6MdDOFEQRcCSigCj+4/mciFUSA
SWS4dMbrpb9FNSIcf9dcLxVM7/6KxgJNfZc9XWzUw77Jg8x92ZdOfVhHOux5IZC+UvSKWB4dyfcI
tE8C3p9bbU9VGyY5vLCAiIb4qQKBgQDLiO24GXrIkswF32YtBBMuVgLGCwU9h9HlO9mKAc2m8Cm1
jUE5IpzRjTedc9I2qiIMUTwtgnw42auSCzbUeYMURPtDqyQ7p6AjMujp9EPemcSVOK9vXYLOPtco
xW9MCOdtV6iPkCN7gOqiZXPRKaFbWADp16p8UAIvS/a5XXk5jwKBgQCKkpHi2EISh1uRkhxljyWC
iDCiK6JBRsMvpLbcOv5dKwP5alo1fmdR5PJaV2qvZSj5CYNpMAy1/EDNTY5OSIJU+OKFmQbyhsbm
rdLNLDL4+TcnT7c62/aHO1ohYaf/VCbRhtLlBfqGoQc7+sAc8vmKkesnF7CqCEKDyF/dhrxYdQKB
gCOiZzzNAapayz1+JcVTwwEid6j9JqNXbBc+Z2YwMi+TOFv/P/hwkX/ypeOXnIUcwOIh/YtGBVAC
DQbsz7LcY1HqXiHKYNWNvXgwwO+oiChjxvEkSdsTTIfnK4VSCvU9BxDbQHjdiNDJbL6oar92UN7V
rBYvChJZF7LvUH4YmVpHAoGAbZ2X7XvoeEO+uZ58/BGKOIGHByHBDiXtzMhdJr15HTYjxK7OgTZm
gK+8zp4L9IbvLGDMJO8vft32XPEWuvI8twCzFH+CsWLQADZMZKSsBasOZ/h1FwhdMgCMcY+Qlzd4
JZKjTSu3i7vhvx6RzdSedXEMNTZWN4qlIx3kR5aHcukCgYA9T+Zrvm1FOseQPbLknn7EqhXIjBaT
P8TTvW/6bdPi23ExzxZn7KOdrfclYRph1LHMpAONv/x2xALIf91UB+v5ohy1oDoasLOgij1houRe
2ERKKdwzOZL9SWq6VTdhr/5G994CK72fy5WhyERbDjUIdHaK3M849JJuf8cSrvSb4g==
-----END RSA PRIVATE KEY-----
```


This command provides only you with the private key. You never see the public key, which is added to your Amazon Web Services account for use when launching instances with this keypair.

Copy this private key and save everything starting with the line "-----BEGIN RSA PRIVATE KEY-----" and ending with the line "-----END RSA PRIVATE KEY----" into a file. How you name the file or where you put it is not terribly important, but it is often useful to include the name of the keypair in the filename: for instance, *id-rsa_mykeypair*. You also need to set permissions on that file so that only you can read it. On Unix-compliant systems such as Linux and Mac OS X, you would enter:

```
$ chmod 400 id_rsa-mykeypair
```

Now you can launch an instance that references your new keypair:

```
$ ec2-run-instances -k mykeypair ami-1fd73376
RESERVATION     r-8a01de64    1234567890123    default
INSTANCE     i-a7d32cc3     ami-1fd73376            pending
        mykeypair         m1.small     2008-10-22T16:40:38+0000
    us-east-1a     aki-a72cf9ce     ari-a52cf9cc
```

You should note that the *-k* option is the name of your keypair, not the name of the file in which you stored your private key.

The output looks much the same as before, except you have a new reservation ID and new instance ID, and the "0" from the previous launch is now replaced with the string `mykeypair` to indicate that the instance was launched with the public key for that keypair.

The public key is assigned to the "root" account on the new instance. In other words, you can theoretically SSH into the box as "root" without typing any login information (it's secure because you have the private key on your local machine that matches up with the public key). In practice, however, you still won't be able to reach it, because the default rules for any security group in EC2 deny all network access to instances in that security group.

NOTE

At this point, those accustomed to the proven best practice of hardening an OS by locking out the root account may be ready to jump through a window. We will discuss mechanisms for removing the need to store any accounts on your AMI and enabling you to disable root login later in Chapter 5.

Security Groups

The last element of the reservation listing in *ec2-describe-instances* (which is `default` in all our examples so far) names the *security group* into which your instance was launched.

The closest real-world analogy to a security group is a network segment protected by a firewall. When created, the firewall comes up with a single rule to deny all traffic from all sources to all destinations on all ports.

For a number of reasons, it's actually a terrible analogy. For the purposes of this chapter, however, the analogy works and we will run with it.

A SECURITY GROUP ISN'T A NETWORK SEGMENT

For the purposes of this chapter, I am using the analogy of a network segment protected by a firewall to describe security groups. Although the analogy works for this chapter, you should most definitely not confuse a security group with a network segment. The following are some of the features of a security group:

- Instances in the same security group do not share an IP block, either in terms of their public or private IP addresses. In fact, they may even exist in different availability zones. As a result, you cannot run applications that rely on multicast or broadcast traffic.
- Although you can put your server in "promiscuous mode," the only network traffic you will ever see is traffic in which your server is an endpoint. This feature significantly increases the security of EC2, but it also makes implementing network intrusion detection systems difficult.
- Your "firewall" rules simply cover how traffic is allowed to route into the group. It does not provide anti-virus protection or other kinds of content-based filtering.
- Instances within the same security group are not allowed to talk to each other unless you create a rule that allows it.
- Once an instance is started, you cannot change its security group membership.

The practical impact is that you cannot access an instance in a new security group in any manner until you grant access using the *ec2-authorize* command:

```
ec2-authorize GROUP_NAME [OPTIONS]
```

To get shell access to your new instance, you will need to open up the SSH port:

```
$ ec2-authorize default -p 22
PERMISSION      default  ALLOWS  tcp     22       22       FROM     CIDR    0.0.0.0/0
```

If this is going to be a web server, you will of course want to add HTTP and HTTPS access:

```
$ ec2-authorize default -p 80
PERMISSION      default  ALLOWS  tcp     80       80       FROM     CIDR    0.0.0.0/0
$ ec2-authorize default -p 443
PERMISSION      default  ALLOWS  tcp     443      443      FROM     CIDR    0.0.0.0/0
```

This output shows that access to the default security group now exists for the specified ports from any place on the Internet (`FROM CIDR 0.0.0.0/0` means any IP address).

WARNING

I opened up SSH to the entire world for this example, but I do not mean to imply that this is actually a good idea. In general, if at all possible, all rules should define a specific source. Obviously, HTTP and HTTPS are generally open to the world, but almost everything else should be closed off completely or selectively allowed.

To narrow access to a particular subnet of incoming traffic, you could have specified the source (`-s`) IP address or subnet from which you and your development team will be accessing the instance:

```
$ ec2-authorize default -P tcp -p 22 -s 10.0.0.1/32
```

That command will provide TCP access to port 22 only from 10.0.0.1. You should now have SSH access to that server:

```
ssh -i PRIVATE_KEY_FILE root@PUBLIC_IP
```

In this case, you could enter:

```
$ ssh -i id_rsa-mykeypair root@ec2-75-101-201-11.compute-1.amazonaws.com
```

You will then be logged in as root on your new instance.

You are not limited to a single security group. You can create additional security groups using the *ec2-add-group* command:

```
ec2-add-group GROUP -d DESCRIPTION
```

For example:

```
$ ec2-add-group mygroup -d MyGroup
GROUP     mygroup  MyGroup
```

Using multiple groups lets you define different rule sets for different kinds of instances—for example, opening HTTP/HTTPS access to a load balancer group while totally closing off access to an application server group, save for Tomcat *modjk* access from the load balancer group.

We will be discussing the setup of security groups for solid web application security in Chapter 5.

Availability Zones

One concept I glossed over earlier was availability zones. An availability zone maps roughly to a physical data center. As I write this book, Amazon provides three availability zones in the United States and two in Western Europe. The most important feature of an availability zone is that any two availability zones have distinct physical infrastructures. Thus the failure of part or all of one availability zone does not impact the other availability zones, except insofar as the other availability zones will then start to take on support for systems formerly operating in the impaired availability zone. Amazon's SLA guarantees 99.95% availability of at least two availability zones within a given region.

In our earlier examples, we saw the availability zone `us-east-1a`. This identifier is not for any particular data center; `us-east-1a` for your Amazon account is probably different than my account's `us-east-1a`.

Understanding availability zones is important for a couple of reasons:

- When you launch one instance in one availability zone and another instance in a second zone, you gain infrastructural redundancy. That is, one instance is almost certain to survive the failure of the other instance, regardless of the reason for the failure.
- You pay for traffic between any two availability zones. In other words, if you have a MySQL master in one availability zone and a slave in another, you pay bandwidth costs for all of the data transfer between the master and slave. You do not pay for that traffic if the two are in the same availability zone. On the other hand, you lose the high-availability benefits of the master/slave setup if they are in the same availability zone.

When you launch an instance without specifying an availability zone, Amazon picks one for you. In general, when launching your first instance, you want Amazon to make the selection for you because it will pick the availability zone with the greatest resource availability. As you add instances into your infrastructure, you will want to specify the target availability zone to maximize redundancy where this is critical and minimize bandwidth charges where redundancy is not critical.

Static IP Addresses

Every time you launch a new instance in EC2, Amazon dynamically assigns it both public and private IP addresses. Later in this book, I will introduce techniques to minimize the need for static IP addresses, but the need can never be eliminated. For example, you almost certainly need a website to have a fixed IP address mapped by your DNS server. Amazon supports this need through elastic IP addresses.

A new Amazon account comes with the ability to create five elastic (static) IP addresses. You are not charged for these IP addresses until you actually create them. Even then, you are charged only for creating them *without* assigning them to a running EC2 instance, or for repeatedly reassigning them. Elastic IP addresses are one of the few things in life you get charged for when you do not use them.

> **NOTE**
>
> **Be a good Internet citizen. Though the charges for IP addresses won't break your bank account, you should allocate an IP address only when you know you are going to need it. The Internet is running out of IP addresses, and the global problem is intensified at the local level because each organization is allocated a fixed number of IP addresses. So, every time people allocate IP addresses that aren't associated with an active instance, it strains the system. Don't say that IPv6 will fix the problem unless you are willing to invest a huge amount of time learning a whole new paradigm, reinstalling new versions of software, and configuring servers from scratch. Amazon, however, does not currently support external routing of IPv6 addresses.**

To allocate new elastic IP address:

```
$ ec2-allocate-address
ADDRESS 67.202.55.255
```

That IP address is now "yours" to assign to an EC2 instance:

```
$ ec2-associate-address -i i-a7d32cc3 67.202.55.255
ADDRESS 67.202.55.255 i-a7d32cc3
```

Finally, you can list all of your allocated addresses:

```
$ ec2-describe-addresses
ADDRESS    67.202.55.255 i-a7d32cc3
ADDRESS    75.101.133.255    i-ba844bc3
```

When you assign an elastic IP address to an instance, the old public address goes away and is replaced by your assigned elastic IP address. If that instance is lost for any reason, you can bring up a new instance and assign the old instance's elastic IP address to it. As a result, you retain public access to your system via a single, static IP address.

The private IP address for an instance always remains the dynamic address assigned to it at launch.

Data Storage in EC2

Amazon provides three very different kinds of storage in its cloud infrastructure:

- Persistent cloud storage
- Ephemeral instance storage

- Elastic block storage

Amazon S3 is the mechanism through which Amazon offers persistent cloud storage.

When you launch a new EC2 instance, it comes with ephemeral instance storage whose lifespan matches the instance it supports. In other words, if you lose the instance or terminate it, you lose everything that was in its ephemeral storage.

The final kind of storage is elastic block storage (EBS). EBS is basically a network-based block storage device, similar to a SAN in a physical infrastructure. You may create volumes varying in size from 1 GB to 1 TB and mount any number of volumes from a single Amazon EC2 instance. You may not, however, share a volume directly between two EC2 instances.

Your gut reaction is almost certainly, "I want door number 3! I want that block storage thing!" Yes, block storage is definitely a good thing. In a well-crafted cloud infrastructure, however, you will make intelligent use of all three kinds of storage, using each where it is best suited to your various storage requirements. Table 2-1 compares the various storage options for data in your EC2 instance.

TABLE 2-1. Comparison of EC2 data storage options

	Amazon S3	**Instance**	**Block storage**
Speed	Low	Unpredictable	High
Reliability	Medium	High	High
Durability	Super high	Super low	High
Flexibility	Low	Medium	High
Complexity	High	Low	High
Cost	Medium	Low	High
Strength	DR management	Transient data	Operational data
Weakness	Operational data	Nontransient data	Lots of small I/O

THE RELIABILITY AND DURABILITY OF S3

S3 has medium reliability but super-high durability because it is simultaneously the most durable of the options but the least reliable. When you put data in S3, you know it will be there tomorrow and the day after and the day after that. No data loss, no corruption. It does not really matter what happens. The problem is that S3 has a pretty weak track record in terms of availability. In the past year, I have seen one entire day without any access to S3 and a number of multihour outages. To my knowledge, however, S3 has never lost any data, and its overall architecture makes it unlikely to lose data. The irony is that, until recently, S3 was the only cloud infrastructure service that Amazon offered with a guaranteed service level.

Another issue is the unpredictability of performance in the instance storage. You might actually think that it should be faster than the other options, and sometimes it is. Sometimes, however, it is unbelievably slow—slower than an NFS mount over a 10bT Ethernet connection. EBS, on the other hand, consistently gives you the performance of a SAN over a GB Ethernet connection.

EBS volume setup

To get started with elastic block storage, create an EBS volume:

```
ec2-create-volume --size SIZE_GB -z ZONE
```

This command creates a block volume of the specified size in the specified availability zone.

One key requirement of a block storage volume is that it must be in the same availability zone as your EC2 instance—you cannot mount volumes across availability zones. Therefore, you will always specify the target availability zone when creating the block volume:

```
$ ec2-create-volume --size 10 -z us-east-1a
VOLUME vol-9a773124 800 creating 2008-10-20T18:21:03+0000
```

You must wait for the volume to become available before you can assign it to an EC2 instance. Check its availability with *ec2-describe-volumes*:

```
$ ec2-describe-volumes vol-9a773124
VOLUME vol-9a773124 800 available 2008-10-20T18:21:03+0000
```

At that point, Amazon starts charging you for the volume, but it is not in use by any instance. To allow an instance to use it, you must attach it to a specific instance in the same availability zone:

```
$ ec2-attach-volume vol-9a773124 -i i-a7d32cc3 -d /dev/sdh
ATTACHMENT vol-9a773124 i-a7d32cc3 /dev/sdh attaching 2008-10-20T18:23:27+0000
```

This command tells EC2 to attach your newly created volume to the specified instance and assign the device */dev/sdh* to the volume (the device name I chose reflects the Unix/Linux device layout, and *sdh* is a common name for a SCSI device on Linux). If the instance lies in a different availability zone or is in use by another instance, you will receive an error message. Otherwise, at this point you have a raw device attached to your instance that you can mount and format in whatever manner you choose.

The most common thing a Linux system does to make a volume usable is format the new volume as an ext3 drive (I prefer XFS because you can freeze the filesystem) and then mount it. To perform these basic tasks, SSH into your instance and execute the following commands:

```
$ mkdir /mnt/myvolume
$ yes | mkfs -t ext3 /dev/sdh
$ mount /dev/sdh /mnt/myvolume
```

You now have a 10 GB volume ready for use by the instance in question. You can use it in any way you would use an ext3 volume.

Volume management

As with any other kind of device, an EBS volume has the potential to become corrupted if not properly disconnected.

> **WARNING**
>
> **You should always, always, *always* unmount an EBS volume before detaching it. If you are running a database engine on that volume, always, always, always shut down the database before unmounting the volume.**

As part of your system shutdown process, your instance should be cleanly unmounting volumes. If, however, you intend to detach a volume outside the normal shutdown process, you should first manually unmount it (and it wouldn't hurt to *sync* it, too):

```
$ umount /mnt/myvolume
```

At that point, it is safe to detach the instance so that it can be available to a different instance:

```
$ ec2-detach-volume vol-9a773124    -i i-a7d32cc3
ATTACHMENT vol-9a773124 i-a7d32cc3 /dev/sdh detaching 2008-10-20T18:55:17+0000
```

You can now attach the volume to another instance.

Snapshots

The ability to take a snapshot of a volume is a particularly fine feature of Amazon's elastic block storage. You can make a snapshot of your volume as often as you like. EC2 automatically saves the snapshot to S3, thus enabling a quick, powerful backup scheme.

> **WARNING**
>
> **Although EBS snapshots are a particularly powerful backup mechanism, keep in mind that they are entirely unportable. You cannot take your EBS snapshots out of the Amazon cloud. Even if you could, you wouldn't be able to make use of them. In Chapter 6, I cover approaches to taking advantage of EBS snapshotting while developing a portable backup strategy.**

Create a snapshot using the *ec2-create-snapshot* command. Before you create a snapshot, however, you will want to make sure the volume you are snapshotting is in a consistent state. In other words, you likely need to stop any write operations on the volume. How you do this depends on the applications writing to the volume.

The biggest concern with filesystem consistency will be any databases stored on the volume. It is absolutely critical that you stop all database write operations before you take your snapshot.

HOW TO LOCK MYSQL FOR A SNAPSHOT

With MySQL, for example, you can either put a lock on the entire engine or put the database in read-only mode. The safest thing is to get a lock on the engine. From the MySQL command-line utility, you can execute the following command:

```
FLUSH TABLES WITH READ LOCK
```

Leave the command line open, lock your filesystem, and create the snapshot. The best filesystem for use with EBS on Linux is XFS, thanks to the *xfs_freeze* command.

Once the snapshot is created, you can close the client, releasing the lock. You need to wait only for the *ec2-create-snapshot* command to return—you do not need to wait for the snapshot to become `completed`.

Just about any database that supports warm backups will adapt well to the EBS snapshot strategy.

To create a snapshot:

```
$ ec2-create-snapshot vol-9a773124
SNAPSHOT snap-21ab4c89b vol-9a773124 pending 2008-10-20T19:02:18+0000
```

Once you create a snapshot, you can immediately begin writing to the volume again; you do not need to wait for the snapshot to complete before resuming writes. The full snapshot process will take time to complete in the background, so the snapshot is not immediately available for use. You can use the *ec2-describe-snapshots* command to identify when the snapshot is ready for use:

```
$ ec2-describe-snapshots snap-21ab4c89b
SNAPSHOT snap-21ab4c89b vol-9a773124 pending 2008-10-20T19:02:33+0000 20%
```

When done, the status field will change from `pending` to `completed` and show the amount completed as `100%`.

The snapshot is not a usable volume itself, but when you need access to the data on it, you can create volumes from it:

```
$ ec2-create-volume --snapshot snap-21ab4c89b -z us-east-1a
VOLUME vol-13b692332a 800 creating 2008-02-15T19:11:36+0000
```

You then have a preformatted volume with all of the data from the old volume ready for use by a new EC2 instance. You can create any number of volumes based on the same snapshot.

This snapshot feature enables you to do a number of powerful things:

- Rapidly create duplicate environments for system testing, load testing, or other purposes that require an exact duplication of production
- Keep running backups that have a minimal impact on your production environment

Another feature of these snapshots is that they are incremental. In other words, if you make one snapshot of a 20 GB volume at 18:00 and another at 18:10, you end up storing in S3 only the original 20 GB plus whatever changed between 18:00 and 18:10. These incremental snapshots thus make it possible to constantly back up your volumes.

AMI Management

When you get started, you will leverage preexisting generic AMIs to bootstrap your efforts. Before you get anywhere near production, however, you will need to build and manage your own library of machine images. Later chapters will discuss in detail strategies for managing machine images, whether for Amazon or another cloud. But to end this chapter, I'll explain the basic procedure you'll need to know to create and register a machine image.

An Amazon machine image contains the root filesystem for your instance. It does not contain anything in your ephemeral storage (files under */mnt* in most setups). To build an AMI, you will need a copy of your Amazon EC2 certificate and private key (the two files ending in *.pem* that you got when setting up your Amazon account). Place those files in */mnt*, because you do not want your Amazon keys embedded in your AMI. You should probably also clean up */tmp* and */var/tmp* so that you are not wasting a bunch of space in S3 on temporary files. Finally, if you are running a database instance on the instance's root partition, stop the database.

Your first task is to bundle up your image:

```
$ cd /mnt
$ sudo mkdir ami
$ sudo ec2-bundle-vol -d /mnt/ami -k /mnt/pk-ZZZ.pem \
      -c /mnt/cert-ZZZ.pem -u 1234567890123 -r i386 -p myami
```

This process will take a long time, and appear to be doing nothing for a while. The command bundles up your root filesystem, breaks it into small parts, and stores it under */mnt/ami/myami*.

Once the process has completed, you will end up with dozens of parts to your AMI, as well as a manifest file called */mnt/ami/myami/myami.manifest.xml*.

At this point, you need to upload the AMI bundle to Amazon S3:

```
$ s3cmd mb s3://myami
$ sudo ec2-upload-bundle -b myami -m /mnt/ami/myami.manifest.xml \
      -a ACCESS_KEY -s SECRET_KEY
```

NOTE

The access key and secret key used to upload your bundle are your S3 access and secret keys, not the EC2 certificates used in creating the bundle.

This command will take some time as it uploads all of the parts into S3. In my experience, this command has a very high likelihood of failure as it is executing a number of S3 PUT commands. If it fails, just try again. Eventually it will succeed.

There's one final step before you can leverage the image: registering the AMI with EC2. To register an AMI:

```
$ ec2-register myami/myami.manifest.xml
IMAGE    ami-33a2d51c
```

When launching new instances, you can use this AMI ID to make them launch from your new instance.

CHAPTER THREE

Before the Move into the Cloud

MOST READERS OF THIS BOOK HAVE BUILT WEB APPLICATIONS deployed in traditional data centers. Now that we have developed a common understanding of what the cloud is and how Amazon implements cloud computing, it's time to look at the concerns you may have when moving into the cloud.

This chapter covers a broad range of these considerations, and I will only touch lightly on approaches to dealing with them at this point. Each of the chapters that follow dives in deeper to solving some of the challenges raised in this chapter.

Know Your Software Licenses

When I cover the issues people face when moving into the cloud, I always start with licensing because it's a nontechnical problem that is too easy to overlook. Just because you have licensing that works for you in a traditional data center does not mean you can port those licenses into the cloud.

With many cloud environments in operation today, you pay for resources by the CPU-hour. The cheapest virtual machine in the Amazon Cloud, for example, costs $0.10 for each hour you leave the instance in operation. If it is up for 10 hours and then shut down, you pay just $1.00—even if that is the only use you make of the Amazon cloud for that month.

In a real world, you might have the following operating scenario:

- From midnight to 9 a.m., run your application on two application servers for redundancy's sake.
- From 9 a.m. to 5 p.m., launch six additional application servers to support business-hour demand.
- For the evening hours through midnight, reduce the system down to four application servers.

Adding all that up, you pay for 110 hours of computing time. If you were using physical servers, you would have to purchase and run eight servers the entire time.

Unfortunately, not all software vendors offer licensing terms that match how you pay for the cloud. Traditional software licenses are often based on the number of CPUs. An organization that uses 10 application servers must pay for 10 application server licenses—even if 5 of them are shut down during the late night hours.

So, when moving to the cloud, you must understand your licensing terms and, in particular:

- Does the software license support usage-based costs (CPU-hour, user, etc.)?
- Does the software allow operation in virtualized environments?

Because the cloud makes it so easy to launch new instances, you can readily find yourself in a situation in which a lower-level staff member has launched instances of software for which you don't have the proper licensing and, as a result, has violated your license agreements.

The ideal cloud-licensing model is open source. In fact, the flexibility of open source licensing has made the Amazon cloud possible. If you can remove licensing concerns altogether from your cloud deployments, you are free to focus on the other challenges of moving to the cloud. Although some open source solutions (such as Apache and most Linux distributions) let you do whatever you want, you may have to deal with licenses if you get supported versions of open source software, such as Red Hat Enterprise Linux or MySQL Enterprise. Luckily, these licensed offerings tend to be cloud-friendly.

Beyond pure open source, the best licensing model for the cloud is one that charges by the CPU-hour. As the cloud catches on, more and more software vendors are offering terms that support hourly license charges. Microsoft, Valtira, Red Hat, Vertica, Sun, and many other companies have adopted per-CPU-hour terms to support cloud computing. Oracle promotes their availability in the cloud but unfortunately retains traditional licensing terms.

Software that offers per-user licensing can work adequately in the cloud as well. The challenge with such software is often how it audits your licensing. You may run the risk of violating your terms, the license may be tied to a specific MAC or IP address, or the software license management system may not be smart enough to support a cloud environment and unreasonably prevent you from scaling it in the cloud.

The worst-case scenario* in the cloud is software that offers per-CPU licensing terms. As with some per-user systems, such software may come with license management tools that make life difficult. For example, you may have to create a custom install for each instance of the software. Doing that impairs the flexibility that the cloud offers.

Some CPU-based software licenses require validation against a licensing server. Any software with this requirement may ultimately be inoperable in the cloud if it is not smart enough to recognize replacement virtual servers on the fly. Even if it can recognize replacements, you'll have to make sure that the license server allows you to start the number of servers you need.

Unless a license server is going to be an impediment, however, the result is no worse than a physical infrastructure. If all of your software fits into this model, the benefits of the cloud may be small to nonexistent.

The Shift to a Cloud Cost Model

As I noted at the start of this chapter, you pay for resources in the cloud as you use them. For Amazon, that model is by the CPU-hour. For other clouds, such as GoGrid, it's by the RAM hour. Let's look at how you can anticipate costs using the example resource demands described earlier (two application servers from midnight until 9 a.m., eight from 9 a.m. until 5 p.m., and four from 5 p.m. until midnight).

Suppose your core infrastructure is:

$0.10/CPU-hour: one load balancer
$0.40/CPU-hour: two application servers
$0.80/CPU-hour: two database servers

Each day you would pay:

$2.40 + $44.00 + $38.40 = $84.80

Your annual hosting costs would come to $30,952.00—not including software licensing fees, cloud infrastructure management tools, or labor.

How to Approach Cost Comparisons

The best way to compare costs in the cloud to other models is to determine the total cost of ownership over the course of your hardware depreciation period. Depending on the organization, a hardware depreciation period is generally two or three years. To get an accurate picture of your total cost of ownership for a cloud environment, you must consider the following cost elements:

* Well, maybe not the worst-case scenario. The worst-case scenario is software that specifically prohibits its use in the cloud or in a virtualized environment.

- Estimated costs of virtual server usage over three years.
- Estimated licensing fees to support virtual server usage over three years.
- Estimated costs for cloud infrastructure management tools, if any, over three years.
- Estimated labor costs for creating machine images, managing the infrastructure, and responding to issues over three years.
- Any third-party setup costs for the environment.

If you want a truly accurate assessment of the three-year cost, you will need to take into account when the costs are incurred during the three-year period and adjust using your organization's cost of capital. This financial mumbo jumbo is necessary only if you are comparing cloud costs against a substantial up-front investment in a purchased infrastructure, but if you understand it, it's still a good idea.

SUNK COSTS AND EXISTING INFRASTRUCTURE

To follow up on the discussion in Chapter 1, in the sidebar "Urquhart on Barriers to Exit" on page 16, this cost analysis ignores any sunk costs. If you can leverage existing infrastructure without incurring any additional costs for leveraging that infrastructure, you should treat that as a $0 cost item. If you have servers sitting around and IT resources with excess availability, you may find that your existing infrastructure is going to have a lower total cost. When making that consideration, however, you should also consider whether delaying your entry into the cloud will have any long-term costs to the organization that offset the savings of leveraging existing infrastructure.

In comparison, you must examine the following elements of your alternatives:

- What are your up-front costs (setup fees, physical space investment, hardware purchases, license fee purchases)?
- What labor is required to set up the infrastructure?
- What are the costs associated with running the infrastructure (hosting space, electricity, insurance)?
- What are the labor costs associated with supporting the hardware and network infrastructure?
- What are the ongoing license subscription/upgrade fees? Maintenance fees?

In Chapter 1, I provided an example comparison for a simple transactional application infrastructure in an internal IT environment against a managed services environment and a cloud environment. Although that example provides a good view of how the cloud can be beneficial, it is absolutely critical you perform a cost analysis using your own real usage numbers. In doing such a comparison, you should make sure that your estimated cloud infrastructure properly supports your operating requirements, that your estimated internal IT costs take into account concerns such as physical space costs and electricity usage, and that your managed services estimates include costs not generally supported by your managed services agreement.

A Sample Cloud ROI Analysis

In Chapter 1, I broke down the costs of a highly theoretical infrastructure without much explanation of their source. Given what we have discussed here, let's perform an ROI analysis of a specific infrastructure that compares building it internally to launching it in the cloud.

> **NOTE**
>
> **Please don't get caught up in the specific costs and purchase options in this analysis. The purpose is not to provide a definitive ROI analysis for the cloud versus internal data center, but instead to outline how to look at the things you must consider when setting up such an analysis. Your analysis should include the purchase decisions you would make in building an infrastructure and the costs of those decisions to your business.**

This particular example assumes that two application servers easily support standard demand. It also assumes, however, that the business has a peak period of traffic on the 15th day of each month that lasts for 24 hours. Serving this peak capacity at the same performance levels as standard capacity requires an extra four servers. This system can still function at peak capacity with only two extra servers—but at the expense of degraded performance.

If you're starting from scratch, you will minimally need the following equipment for your IT shop:

- Half rack at a reliable ISP with sufficient bandwidth to support your needs
- Two good firewalls
- One hardware load balancer
- Two good GB Ethernet switches
- Six solid, commodity business servers (we will accept degraded performance on the 15th day)

For the cloud option, you will need just a few virtual instances:

- One medium 32-bit instance
- Four large 64-bit instances during standard usage, scaled to meet peak demand at 8

In addition, you need software and services. Assuming an entirely open source environment, your software and services costs will consist of time to set up the environments, monitoring services, support contracts, and labor for managing the environment. Table 3-1 lays out all of the expected up-front and ongoing costs.

TABLE 3-1. Costs associated with different infrastructures

	Internal (initial)	Cloud (initial)	Internal (monthly)	Cloud (monthly)
Rack	$3,000	$0	$500	$0
Switches	$2,000	$0	$0	$0
Load balancer	$20,000	$0	$0	$73
Servers	$24,000	$0	$0	$1,206
Firewalls	$3,000	$0	$0	$0
24/7 support	$0	$0	$0	$400
Management software	$0	$0	$100	$730
Expected labor	$1,200	$1,200	$1,200	$600
Degraded performance[a]	$0	$0	$100	$0
Totals	**$53,200**	**$1,200**	**$1,900**	**$3,009**

[a] Remember, we have opted to buy only four application servers in the internal infrastructure instead of six. That degraded performance has some cost to the business. This cost is typically lost business or a delay in receiving cash. I picked $100 out of the air here because anything more than about $110 would justify the addition of new servers into the mix and anything less would raise the question of why you would bother building out support for peak capacity at all.

To complete the analysis, you need to understand your organization's *depreciation schedule* and *cost of capital*. For hardware, the depreciation schedule is typically two or three years. For the purposes of this example, I will use the more conservative three-year schedule. This schedule essentially defines the expected lifetime of the hardware and frames how you combine monthly costs with one-time costs.

The cost of capital for most organizations lies in the 10% to 20% range. In theory, the cost of capital represents what your money is worth to you if you were to invest it somewhere else. For example, if you have $10,000, a 10% cost of capital essentially says that you know you can grow that $10,000 to $11,046.69 (with the cost of capital being compounded monthly) after a year through one or more standard investment mechanisms. Another way to look at it is that gaining access to $10,000 in capital will cost you $11,046.69 after a year. Either way, the cost of $10,000 to your business is 10%, calculated monthly.

Ignoring the cost of capital, the internal build-out costs you $121,600 and the cloud costs you $109,254. These numbers result from adding the up-front costs together with the monthly costs after the depreciation period—in this case, three years. In this example, it is obvious that the cloud comes out better, even when we ignore the cost of capital. The cost of capital will always be higher for the option that requires more money earlier. As a result, even if the two numbers had come out even at $100,000, we would know from the schedule for acquiring capital that the cloud would be the more financially attractive option.

If you want to compare scenarios in which the cloud looks slightly more expensive, or if you want to know the level of business you must generate to justify your infrastructure, you must take into account the cost of capital so you can understand each option's true cost to the business. The financial expression is that you want to know the *present value* (the value of those payments to you today if you were forced to pay for it all today) of all of your cash outflows over the course of the depreciation period.

Learning how to calculate present value is a full chapter in most financial textbooks and thus beyond the scope of a book on cloud computing. Fortunately, Microsoft Excel, Apple Numbers, your financial calculator, and just about any other program on earth with financial functions will take care of the problem for you.

For each option, calculate the present value of your monthly payments and add in any up-front costs:

> Internal: = (–PV(10%/12,36,1900,0)) + 53200 = $112,083.34
> Cloud: = (–PV(10%/12,36,3900,0)) + 1200 = $94,452.63

Not only is the cloud cheaper, but the payment structure of the cloud versus up-front investment saves you several thousand dollars.

The final point to note is that you can use the $112,083.34 and $94,452.63 numbers to help you understand how much money these systems need to generate over three years in order to be profitable (make sure that calculation is also using today's dollars).

Where the Cloud Saves Money

As you engage in your cost analysis, you will find merely moderate savings if your application has a fairly static demand pattern. In other words, if you always have the same steady amount of usage, you won't see most of the key cost benefits of being in the cloud. You should see some savings over hosting things in your own infrastructure due to the reduced labor requirements of the cloud, but you might even pay more than some managed services environments—especially if you have true high-availability requirements.

Cost savings in the cloud become significant, and even reach absurd levels, as your variance increases between peak and average capacity and between average and low capacity. My company has an extreme example of a customer that has very few unique visitors each day for most of the year. For about 15 minutes each quarter, however, they have the equivalent of nearly 10 million unique visitors each month hitting the website. Obviously, purchasing hardware to support their peak usage (in other words, for 1 hour per year) would be insanely wasteful. Nevertheless, they don't have the option of operating at a degraded level for that one hour each year. The cloud is the perfect solution for them, as they can operate with a very baseline infrastructure for most of the year and scale up for a few minutes each quarter to support their peak demand.

A common and relatively straightforward set of cost savings lies in the management of nonproduction environments—staging, testing, development, etc. An organization generally requires these environments to be up at certain points in the application development cycle and then torn down again. Furthermore, testing requirements may demand a full duplication of the production environment. In the cloud, you can replicate your entire production infrastructure for a few days of testing and then turn it off.

Service Levels for Cloud Applications

When a company offers a service—whether in the cloud or in a traditional data center—that company generally provides its customers with a *service level agreement* (SLA) that identifies key metrics (*service levels*) that the customer can reasonably expect from the service. The ability to understand and to fully trust the *availability*, *reliability*, and *performance* of the cloud is the key conceptual block for many technologists interested in moving into the cloud.

Availability

Availability describes how often a service can be used over a defined period of time. For example, if a website is accessible to the general public for 710 hours out of a 720-hour month, we say it has a 98.6% availability rating for that month.

Although 98.6% may sound good, an acceptable value depends heavily on the application being measured—or even what features of the application are available. If, for example, Google's spider is down for 24 hours but you can still search and get results, would you consider Google to be down?

Most people consider a system to have *high availability* if it has a documented expectation of 99.99% to 99.999% availability. At 99.999% availability, the system can be inaccessible for at most five minutes and 15 seconds over the course of an entire year.

LUCK IS NOT HIGH AVAILABILITY

Referring to a system as a high-availability system describes what you can expect in terms of future uptime—not what you happened to get one year when you were lucky. The fact that your personal website hosted on an old 486 in your basement has managed to stay up all year does not suddenly make it a high-availability system. If that 486 has roughly a 40% shot of going down within a calendar year, it is completely unreasonable for you to claim it as a high-availability environment, regardless of its past performance.

How to estimate the availability of your system

Most service outages are the result of misbehaving equipment. These outages can be prolonged by misdiagnosis of the problem and other mistakes in responding to the outage in question. Determining expected availability thus involves two variables:

- The likelihood you will encounter a failure in the system during the measurement period.
- How much downtime you would expect in the event the system fails.

The mathematic formulation of the availability of a component is:

$$a = (p - (c \times d))/p$$

where:

a = expected availability
c = the % of likelihood that you will encounter a server loss in a given period
d = expected downtime from the loss of the server
p = the measurement period

So, if your 486 has a 40% chance of failure and you will be down for 24 hours, your 486 uptime is (8760 – (40%×24))/8760, or just shy of 99.9%.

99.9% availability sounds pretty great for an old 486, right? Well, I certainly oversimplified the situation. Do you honestly expect your DSL or cable never to go down? And do you think you can go get a replacement server, configure it, and execute a recovery from backup within 24 hours? What about your networking equipment? How reliable are your backups?

To achieve an accurate availability rating, you need to rate all of the points of failure of the system and add them together. The availability of a system is the total time of a period minus the sum of all expected downtime during that period, all divided by the total time in the period:

$$a = (p - \mathrm{SUM}(c1 \times d1 : cn \times dn))/p$$

If your cable provider generally experiences two outages each year lasting two hours each, the Internet connection's availability is:

$$(8760 - (200\% \times 2))/8760 = 99.95\%$$

Thus, your overall availability is:

$$(8760 - ((40\% \times 24) + (200\% \times 2)))/8760 = 99.84\%$$

This example illustrates the ugly truth about software systems: the more points of failure in the system, the lower its availability rating will be. Furthermore, the amount of time you are down when you go down has a much more profound impact than how often you might go down.

Redundancy mitigates this problem. When you have two or more physical components representing a logical component, the expected downtime of the logical component is the expected amount of time you would expect all of the physical components to be down simultaneously. In other words, the *c*×*d* formula for downtime calculation becomes slightly more complex:

$$(c \times d^n)/(p^{(n-1)})$$

where *n* is the level of redundancy in the system. As a result, when *n* = 1, the formula reduces as you would expect to its simpler form:

$$(c \times d^n)/(p^{(n-1)}) = (c \times d)/(p^0) = c \times d$$

The redundancy of two 486 boxes—and this requires seamless failover to be considered one logical component—now provides a much better availability rating:

$$(8760 - ((40\% \times (24^2))/(8760^{(2-1)}))/8760 = 99.999\%$$

What constitutes availability?

At some level, high availability lies in the eye of the beholder. If availability is a requirement for your system, you need to define not simply the percentage of time it will be available, but what it means to be available. In particular, you need to define the following availability criteria:

- What features must be accessible in order for the system to qualify as available? In the Google example earlier, we would consider Google to be available as long as you can search the site and get results; what the spider is doing is irrelevant.
- Should you include planned downtime? Including planned downtime is somewhat controversial, but you may want to do it for some types of availability measurement and not for others. For example, if your architecture is fully capable of supporting 99.999% availability but you prefer to take the environment down 1 hour each week for system maintenance (due to overzealous security procedures or whatever criterion not related to the theoretical availability of the architecture itself), you could advertise the environment as having 99.999% availability. If, however, you are trying to communicate to end users what they should expect in the way of uptime, saying such an environment has 99.999% availability is misleading.
- What percentage of the time will your environment remain available?

A website, for example, might state its availability requirements as the home page and all child pages will be accessible from outside the corporate network between the hours of 7 a.m. and 9 p.m. 99.999% of the time.

Cloud service availability

As we put together an architecture for deploying web applications in the cloud, we need to understand what we can expect from the cloud. Although the same high-availability concepts apply in the cloud as in a traditional data center, what you might consider a reliable component differs dramatically between the two paradigms.

Perhaps the most jarring idea is that many failures you might consider rare in a traditional data center are common in the cloud. This apparent lack of reliability is balanced by the fact that many of the failures you might consider catastrophic are mundane in the cloud.

For example, an EC2 instance is utterly unreliable when compared to the expected availability of a low-end server with component redundancies. It is very rare to lose such a physical server with no warning at all. Instead, one component will typically fail (or even warn you that it is about to fail) and is replaced by a redundant component that enables you to recover without any downtime. In many other clouds, such as Rackspace and GoGrid, you see the same levels of reliability. In the Amazon cloud, however, your instances will eventually fail without warning. It is a 100% guaranteed certainty.

Furthermore, a failure of a virtual instance is the physical equivalent to a grenade going off in your server with no collateral damage. In other words, the server is completely lost. You won't recover anything

Am I scaring you? This distinction is one of the first stumbling blocks that makes technologists nervous about the cloud. It's scary because in the physical world, the loss of a server is a minor catastrophe. In the virtual world, it's almost a nonevent. In fact, in the cloud, you can lose an entire availability zone and shrug the incident off. In contrast, what would happen to your systems if your entire data center were to suddenly disappear?

In Chapters 4 and 6, I discuss techniques that will not only alleviate any fears I might have created just now, but also will enable you to feel more confident about your cloud environment than you would feel about a physical environment.

Amazon Web Services service levels

One way in which competitors have elected to compete against Amazon is in the area of service levels. Most competitors offer strong service levels in the cloud. Although Amazon has provided its customers with an SLA for S3 for some time, it has only recently added a formal SLA to EC2. The S3 service level promises that S3 will respond to service requests 99.5% of the time in each calendar month. EC2, on the other hand, defines a more complex availability service level. In particular, EC2 promises 99.95% availability of at least two availability zones within a region.

These service levels don't immediately translate into something you can promise your own customers when deploying in the cloud. In particular:

- You need S3 to be available to launch an EC2 instance. If S3 has a 99.5% availability, you will be able to launch new EC2 instances only 99.5% of the time, regardless of how well EC2 is living up to—or exceeding—its promises. This drawback also applies to snapshots and creating volumes because you cannot take a snapshot or create a volume from a snapshot if S3 is down.

- EC2 is living up to its service level as long as two availability zones in the same region are available 99.95% of the time. EC2 can technically live up to this service level with entire availability zones constantly going down.
- You need to architect your application to be able to reliably support the demands on it.

CAN YOU TRUST IT?

It's one thing for Amazon to promise a certain level of availability; what matters is that they live up to that promise.

It is critical here to distinguish between EC2 and S3. EC2 is based on a known technology (Xen) with customizations solving a well-understood problem: virtualization. S3, on the other hand, is a largely homegrown system doing something fairly unique. In my experience, problems with EC2 have been related to fairly mundane data center issues, whereas problems with S3 have involved the more esoteric elements of their proprietary cloud storage software.

Boiling it down to the essence, EC2 is better understood than S3. Amazon is thus more likely to be "right" about its promises relating to EC2 than it is with respect to S3. Reality has borne out this analysis: S3 has had significant outages[†] in the past year and thus failed to live up to the 99.5% availability on occasion. During the same period, I am not aware of EC2 having any trouble living up to 99.95% availability, even though EC2 did not yet have the SLA in place and was labeled a "beta" service.

The key takeaway? You need to build an application infrastructure that does not operationally rely on S3 to meet its availability objectives. I'll help you do that as this book progresses.

Expected availability in the cloud

The key differentiator between downtime in the cloud and downtime in a physical environment lies in how much easier it is to create an infrastructure that will recover rapidly when something negative happens. In other words, even though a physical server is more reliable than a virtual server in the cloud, the cloud enables you to inexpensively create redundancies that span data centers and more quickly recover when a downtime occurs.

Let's compare a very simple setup that includes a single load balancer, two application servers, and a database engine.

† I don't mean to overplay these outages. In my experience, Amazon's communications and transparency during outages have been outstanding, and they are very obviously working hard to improve the reliability for an area in which they are blazing the trail. By the time you read this book, it is very possible that these outages will be quaint relics of Amazon's early growing pains.

When implementing this architecture in a physical data center, you will typically locate this infrastructure in a single rack at a hosting provider using fairly mundane servers with component redundancy and hardware load balancing. Ignoring network failure rates and the possibility of a data center catastrophe, you will probably achieve availability that looks like these calculations:

Load balancer
: 99.999% (people use hardware load balancers partly to minimize component failure)

Application server
: $(8760 - ((30\% \times (24^2))/8760))/8760 = 99.999\%$

Database server
: $(8760 - (24 \times 30\%))/8760 = 99.92\%$

Overall system
: $(8760 - ((24 \times 30\%) + (24 \times (((30\% \times (24^2))/8760)) + (24 \times 30\%)))/8760 = 99.84\%$

For the purposes of this calculation, I assumed that the load balancer is virtually assured of never going down during the course of its depreciation period, that the servers are guaranteed to eventually go down, and that the loss of any of the components will cause 24 hours of downtime.

The best way to improve on this infrastructure cheaply is to have spare parts lying around. By cutting down on the downtime associated with the loss of any given component, you dramatically increase your availability rating. In fact, if you are using a managed services provider, they will generally have extra components and extra servers available that will cut your downtime to the amount of time it takes to swap in the replacement and configure it. On the other hand, if you are managing your servers yourself and you don't have a 24-hour turnaround guarantee from your hardware vendor, your environment may have a truly miserable availability rating.

WHAT DO THESE NUMBERS REALLY MEAN?

Reasonable people can—and will—disagree with the numbers I have used in this section in both directions. The purpose of these calculations is not to make the claim that a traditional data center has a 99.84% availability for this architecture and the cloud has a 99.994% availability. If you read this section that way and quote me on it, I will be very unhappy!

What you should see in these calculations is a process for thinking about availability, and an understanding of how stability in the cloud differs from stability in a physical infrastructure:

- EC2 instances are much less stable than physical servers.
- The reliance on multiple availability zones can significantly mitigate the lack of stability in EC2 instances.

- The lack of stability of a software load balancer is largely irrelevant thanks to the ability to quickly and automatically replace it.

In the cloud, the calculation is quite different. The load balancer is simply an individual instance, and individual server instances are much less reliable. On the other hand, the ability to span availability zones increases clustering availability. Furthermore, the downtime for these nodes is much less:‡

Load balancer

$(8760 - (.17\times80\%))/8760 = 99.998\%$

Application server

$(8760 - (17\%\times((.17^2)/8760)))/8760 = 99.9999\%$

Database server

$(8760 - (.5\times80\%))/8760 = 99.995\%$

Overall system

$(8760 - ((.17\times80\%) + (17\%\times((.17^2)/8760)) + (.5\times80\%)))/8760 = 99.994\%$

This calculation assumes you are using tools to perform automated recovery of your cloud infrastructure. If you are doing everything manually, you must take into account the increased downtimes associated with manual cloud management. Finally, I ignored the impact of S3's lack of reliability. In reality, there is a small chance that when you need to launch a replacement instance (which keeps your downtime to a minimum), you will be unable to do so because S3 is unavailable.

Reliability

Reliability is often related to availability, but it's a slightly different concept. Specifically, reliability refers to how well you can trust a system to protect data integrity and execute its transactions. The instability associated with low availability often has the side effect of making people not trust that their last request actually executed, which can cause data corruption in relational database systems.

> **NOTE**
>
> A system that is frequently not available is clearly not reliable. A highly available system, however, can still be unreliable if you don't trust the data it presents. This could be the case, for instance, if some process or component silently fails.

‡ I am ignoring the impact on performance of losing a particular node and focusing entirely on availability. If your redundant application servers are operating at full capacity, the loss of one of them will result in either degraded performance or loss of availability.

Much of the reliability of your system depends on how you write the code that runs it. The cloud presents a few issues outside the scope of your application code that can impact your system's reliability. Within the cloud, the most significant of these issues is how you manage *persistent data*.

Because virtual instances tend to have lower availability than their physical counterparts, the chance for data corruption is higher in the cloud than it is in a traditional data center. In particular, any time you lose a server, the following factors become real concerns:

- You will lose any data stored on that instance that has not been backed up somewhere.
- Your block storage devices have a chance of becoming corrupted (just as they would in a traditional data center).

I will discuss in detail strategies for dealing with these concerns in Chapter 4. For now, you should remember two rules of thumb for dealing with reliability concerns in the cloud:

- Don't store EC2 persistent data in an instance's ephemeral mounts (not necessarily applicable to other clouds).
- Snapshot your block volumes regularly.

Performance

For the most part, the things you worry about when performing a high-performance transactional application for deployment in a physical data center applies to deployment in the cloud. Standard best practices apply:

- Design your application so logic can be spread across multiple servers.
- If you are not clustering your database, segment database access so database reads can run against slaves while writes execute against the master.
- Leverage the threading and/or process forking capabilities of your underlying platform to take advantage of as much of each individual CPU core as possible.

Clustering versus independent nodes

Depending on the nature of your application, your choke points may be at the application server layer or the database layer. At the application layer, you essentially have two options for spreading processing out across multiple application servers:

- Use a load balancer to automatically split sessions across an indeterminate number of independent nodes. Under this approach, each virtual server instance is entirely ignorant of other instances. Therefore, each instance cannot contain any critical state information beyond the immediate requests it is processing.
- Use a load balancer to route traffic to nodes within a clustered application server. Through clustering, application server nodes communicate with each other to share state

information. Clustering has the advantage of enabling you to keep state information within the application server tier; it has the disadvantage of being more complex and ultimately limiting your long-term scalability. Another challenge with true clustering is that many clustering architectures rely on multicasting—something not available in Amazon EC2 (but available with GoGrid and Rackspace).

I am a strong proponent of the independent nodes approach. Although it can take some skill to architect an application that will work using independent nodes, this architecture has the benefit of being massively scalable. The trick is to make sure that your application state is managed via the shared database, message queue, or other centralized data storage and that all nodes are capable of picking up changes in the state in near real time.

EC2 performance constraints

EC2 systems, in general, perform well. When you select a particular EC2 machine, the CPU power and RAM execute more or less as you would expect from a physical server. Network speeds are outstanding. Differences show up in network and disk I/O performance.

> **WARNING**
>
> **EC2 32-bit instances are almost unbearably slow. This slowness is not so much an artifact of EC2 as it is a reflection of the amount of CPU you get for $0.10/CPU-hour. They are nevertheless useful for scenarios in which performance does not matter, but cost does: prototyping, quick development, operations with low CPU requirements (such as load balancing), etc.**

The three different kinds of data storage have three very different performance profiles:

- Block storage has exactly the kind of performance you would expect for a SAN with other applications competing for its use across a GB Ethernet connection. It has the most reliable performance of any of the options, and that performance is reliably solid.
- S3 is slow, slow, slow (relatively speaking). Applications should not rely on S3 for real-time access.
- Local storage is entirely unpredictable. In general, first writes to a block of local storage are slower than subsequent writes. I have encountered outstanding disk I/O rates as well as I/O rates worse than mounting a WebDAV drive over a 56K modem. I have not seen any pattern that would help me provide an explanation for how to optimize performance. If you start seeing terrible performance, an instance reboot[§] generally seems to take care of the problem (and you should take this into account with your uptime estimates).

[§] An EC2 instance retains its state across reboots. Rebooting is not like terminating or losing the instance.

EC2 disk performance serves the needs of most transactional web applications. If, however, you have an application where disk I/O really matters, you will want to benchmark your application in the EC2 environment before committing to deploying it in EC2.

Security

One of the items that critics of clouds repeatedly hammer on is cloud security. It seems that having your data in the cloud on machines you do not control is very emotionally challenging to people. It also introduces real regulatory and standards compliance issues that you need to consider. In reality, the cloud can be made as secure as—or even more secure than—a traditional data center. The way you approach information security, however, is radically different.

A move into the cloud requires consideration of a number of critical security issues:

- Legal implications, regulatory compliance, and standards compliance issues are different in the cloud.
- There is no perimeter in the Amazon cloud; a security policy focused on perimeter security will not work with Amazon and should not be your focus, even with clouds that support traditional perimeter security.
- Although there have been no publicized exploits, cloud data storage should assume a high-risk profile.
- Virtualization technologies such as Xen may ultimately have their own vulnerabilities and thus introduce new attack vectors.

Legal, Regulatory, and Standards Implications

Unfortunately, the law and standards bodies are a bit behind the times when it comes to virtualization. Many laws and standards assume that any given server is a physically distinct entity. Most of the time, the difference between a physical server and a virtual server is not important to the spirit of a given law or standard, but the law or standard may nevertheless specify a physical server because the concept of virtual servers was not common when the specification was being written.

Before moving into the cloud, you must fully understand all of the relevant laws and standards to which your applications and infrastructure are bound. In all probability, a cloud equivalent will support the spirit of those regulations and standards, but you may have to talk to experts in your specific requirements to find a definitive answer as to whether the cloud can be considered conformant. In cases where a pure cloud infrastructure won't support your needs, you can often craft a mixed infrastructure that will still provide you with many of the benefits of the cloud while still clearly complying with standards and regulations.

WARNING

If you are looking for this book to answer the question, "Is the cloud compliant with specification X?", you will be disappointed. Amazon is working on SAS 70 compliance, and a number of other cloud providers have environments that support different specifications. You will still have to manage your own instances to your target specification. I cannot answer compliance questions. Only a lawsuit or formal certification of your environment will definitively answer them.

Beyond compliance issues, the cloud also introduces legal issues related to where your data is stored:

- Your data may be exposed to subpoenas and other legal procedures that cite your cloud provider instead of citing you, and that may invoke different rights than procedures involving you directly.
- Some nations (in particular, EU nations) have strict regulations governing the physical location of private data.

There Is No Perimeter in the Cloud

The Amazon cloud has no concept of a network segment controlled by firewalls and provides you with minimal ability to monitor network traffic. Each instance sees only the network traffic it is sending or receiving, and the initial policy of the default security group in EC2 allows no incoming traffic at all.

NOTE

Other clouds support the concept of virtual LANs and provide traditional perimeter security. It's a good practice, however, not to rely on perimeter security as your primary approach to system security, regardless of what cloud you are using.

In short, you don't set up a firewall and a network IDS tool to secure your cloud infrastructure (I know—I have trivialized perimeter security in the physical world). In Chapter 5, I cover how you can effectively secure traffic in and out of your virtual servers without access to traditional firewalls and network perimeters.

The Risk Profile for S3 and Other Cloud Storage Solutions Is Unproven

While the lack of traditional perimeter security makes security experts uncomfortable, the placement of critical data assets into a data center that is under someone else's control really sends some cloud evaluators off the deep end.

You never know where your data is when it's in the cloud. However, you know some basic parameters:

- Your data lies within a Xen virtual machine guest operating system or EBS volume, and you control the mechanisms for access to that data.
- Network traffic exchanging data between instances is not visible to other virtual hosts.
- S3 storage lies in a public namespace but is accessible through objects that are private by default.
- Amazon zeros out all ephemeral storage between uses.

You should therefore make the following assumptions about your data:

- Except for the possibility of virtualization-specific exploits, data within a virtual machine instance is basically as secure as data sitting on a physical machine. If you are truly paranoid, you can leverage encrypted filesystems.
- Your S3 objects are inherently insecure and any sensitive data should definitely be encrypted with a strong encryption option before being placed in S3.
- Your block storage snapshots are probably reasonably secure. Though they are stored in S3, they are not accessible via normal S3 access protocols. Nevertheless, if you are truly paranoid, it would not hurt to encrypt your block storage volumes.
- You need to make sure that you retain copies of your data outside of your cloud provider in case your cloud provider goes bankrupt or suffers some other business interruption.

Disaster Recovery

Disaster recovery is the art of being able to resume normal systems operations when faced with a disaster scenario. What constitutes a disaster depends on your context. In general, I consider a disaster to be an anomalous event that causes the interruption of normal operations. In a traditional data center, for example, the loss of a hard drive is not a disaster scenario, because it is more or less an expected event. A fire in the data center, on the other hand, is an abnormal event likely to cause an interruption of normal operations.

The total and sudden loss of a complete server, which you might consider a disaster in a physical data center, happens—relatively speaking—all of the time in the cloud. Although such a frequency demotes such events from the realm of disaster recovery, you still need solid disaster recovery processes to deal with them. As a result, disaster recovery is not simply a good idea that you can keep putting off in favor of other priorities—it is a requirement.

What makes disaster recovery so problematic in a physical environment is the amount of manual labor required to prepare for and execute a disaster recovery plan. Furthermore, fully testing your processes and procedures is often very difficult. Too many organizations have a disaster recovery plan that has never actually been tested in an environment that sufficiently replicates real-world conditions to give them the confidence that the plan will work.

Disaster recovery in the cloud can be much more automatic. In fact, some cloud infrastructure management tools will even automatically execute a disaster recovery plan without human intervention.

What would happen if you lost your entire data center under a traditional IT infrastructure? Hopefully, you have a great off-site backup strategy that would enable you to get going in another data center in a few weeks. While we aren't quite there yet, the cloud will soon enable you to move an entire infrastructure from one cloud provider to another and even have that move occur automatically in response to a catastrophic event.

Another advantage for the cloud here is the cost associated with a response to a disaster of that level. Your recovery costs in the cloud are almost negligible beyond normal operations. With a traditional data center, you must shell out new capital costs for a new infrastructure and then make insurance claims. In addition, you can actually test out different disaster scenarios in the cloud in ways that are simply unimaginable in a traditional environment.

I've covered the key concerns that affect the architecture of your applications and server configuration as you consider moving into the cloud. The rest of this book guides you along the path.

CHAPTER FOUR

Ready for the Cloud

To a large extent, one secret will guide you in deploying an application capable of leveraging all of the benefits of the cloud into the cloud:

> *Do what you would do anyway to build a highly scalable web application.*

In the absence of specific regulatory or standards compliance issues, if your application can run behind a load balancer across a number of application server nodes without any problems, you are pretty much all set. This chapter will help you determine how to move that application into the cloud.

On the other hand, many web applications have been built with a single server in mind, and their creators aren't sure whether they can safely move to a clustered environment. If you fit into this category—or if you know for a fact your application won't work in a clustered environment—this chapter will tell you what you need to make your application ready for the cloud.

Web Application Design

I cannot possibly address the intricacies of each of the many platforms developers use in building web applications, but most of the issues you will face have nothing to do with your underlying choice of platform. Whether written in .NET, Ruby, Java, PHP, or anything else, web applications share a similar general architecture—and architecture makes or breaks an application in the cloud.

Figure 4-1 illustrates the generic application architecture that web applications share.

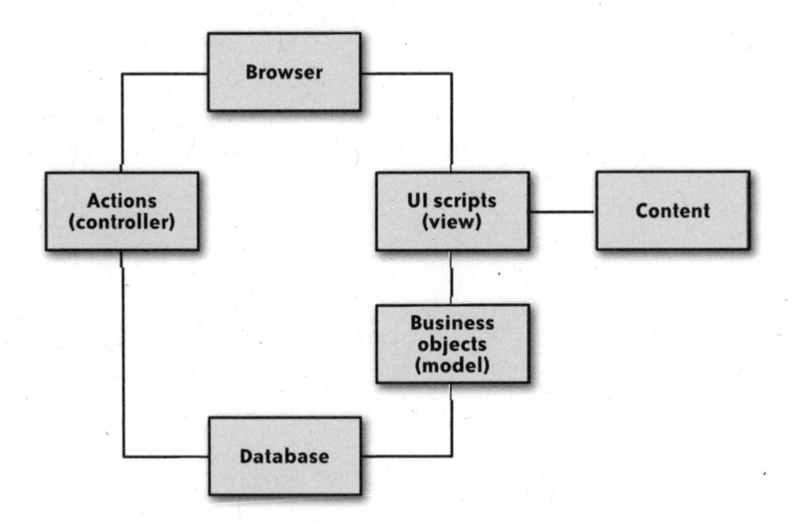

FIGURE 4-1. Most web applications share the same basic architecture

You may move around or combine the boxes a bit, but you are certain to have some kind of (most often scripting) language that generates content from a combination of templates and data pulled from a model backed by a database. The system updates the model through actions that execute transactions against the model.

System State and Protecting Transactions

The defining issue in moving to the cloud is how your application manages its state. Let's look at the problem of booking a room in a hotel.

The architecture from Figure 4-1 suggests that you have represented the room and the hotel in a model. For the purposes of this discussion, it does not matter whether you have a tight separation between model, view, and data, or have mixed them to some degree. The key point is that there is some representation of the hotel and room data in your application space that mirrors their respective states in the database.

How does the application state in the application tier change between the time the user makes the request and the transaction is changed in the database?

The process might look something like this basic sequence:

1. Lock the data associated with the room.
2. Check the room to see whether it is currently available.

3. If currently available, mark it as "booked" and thus no longer available.
4. Release the lock.

The problem with memory locks

You can implement this logic in many different ways, not all of which will succeed in the cloud. A common Java approach that works well in a single-server environment but fails in a multiserver context might use the following code:

```
public void book(Customer customer, Room room, Date[] days)
    throws BookingException {
    synchronized( room ) { // synchronized "locks" the room object
        if( !room.isAvailable(days) ) {
            throw new BookingException("Room unavailable.");
        }
        room.book(customer, days);
    }
}
```

Because the code uses the Java locking keyword `synchronized`, no other threads in the current process can make changes to the room object.* If you are on a single server, this code will work under any load supported by the server. Unfortunately, it will fail miserably in a multiserver context.

The problem with this example is the memory-based lock that the application grabs. If you had two clients making two separate booking requests against the same server, Java would allow only one of them to execute the `synchronized` block at a time. As a result, you would not end up with a double booking.

On the other hand, if you had each customer making a request against different servers (or even distinct processes on the same server), the synchronized blocks on each server could execute concurrently. As a result, the first customer to reach the `room.book()` call would lose his reservation because it would be overwritten by the second. Figure 4-2 illustrates the double-booking problem.

The non-Java way of expressing the problem is that if your transactional logic uses memory-based locking to protect the integrity of a transaction, that transaction will fail in a multiserver environment—and thus it won't be able to take advantage of the cloud's ability to dynamically scale application processing.

One way around this problem is to use clustering technologies or cross-server shared memory systems. Another way to approach the problem is to treat the database as the authority on the state of your system.

* I beg Java programmers to forgive my oversimplification and technically incorrect description of the `synchronized` keyword. I intend the explanation to help readers in general understand what is happening in this context and not to teach people about multithreading in Java.

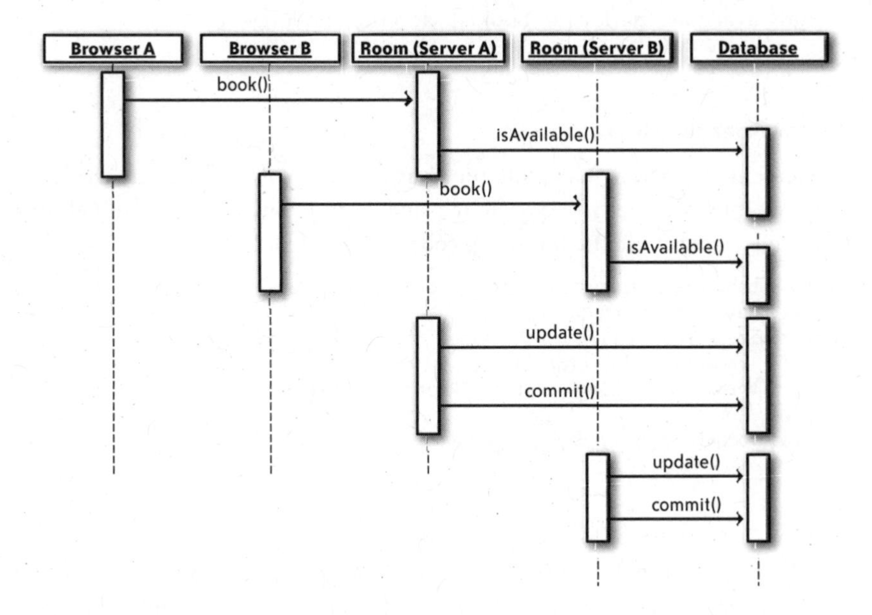

FIGURE 4-2. The second client overwrites the first, causing a double-booking

---

WHAT IF MY APPLICATION USES MEMORY LOCKS?

I can hear a lot of you—especially those of you who have massively multithreaded applications—cursing me right now. If you find yourself in a situation in which you use memory-based locking and reworking the application away from that model is impractical, you can still move into the cloud. You simply won't be able to scale your application across multiple application servers.

The way around this concern is to lock everything using a shared locking mechanism—typically your database engine. The only other common alternative is to write some kind of home-grown distributed transaction management system. But why do that when your database already has one?

Transactional integrity through stored procedures

I am not a huge fan of stored procedures. A key benefit of stored procedures, however, is that they enable you to leverage the database to manage the integrity of your transactions. After all, data integrity is the main job of your database engine!

Instead of doing all of the booking logic in Java, you could leverage a MySQL stored procedure:

```
DELIMITER |

CREATE PROCEDURE book
```

```
(
  IN customerId BIGINT,
  IN roomId BIGINT,
  IN startDate DATE,
  IN endDate DATE,
  OUT success CHAR(1)
)
BEGIN
  DECLARE n DATE;
  DECLARE cust BIGINT;
  SET success = 'Y';
  SET n = startDate;

  bookingAttempt:
  REPEAT
    SELECT customer INTO cust FROM booking
    WHERE room_id = roomId AND booking_date = n;
    IF cust IS NOT NULL AND cust <> customerId
    THEN
      SET success = 'N';
      LEAVE bookingAttempt;
    END IF;
    UPDATE booking SET customer = customerId
    WHERE room_id = roomId AND booking_date = n;
    SET n = DATE_ADD(n, INTERVAL 1 DAY);
  UNTIL n > endDate
  END REPEAT;
  IF success = 'Y' THEN
    COMMIT;
  ELSE
    ROLLBACK;
  END IF;
END
|
```

This method goes through each row of the booking table in your MySQL database and marks it booked by the specified customer. If it encounters a date when the room is already booked, the transaction fails and rolls back.

An example using the stored procedure follows, using Python:

```
def book(customerId, roomId, startDate, endDate):
    conn = getConnection();
    c = conn.cursor();
    c.execute("CALL book(%s, %s, %s, %s, @success)", \
              (customerId, roomId, startDate, endDate));
    c.execute("SELECT @success");
    row = c.fetchone();
    success = row[0];
    if success == "Y":
      return 1
    else:
      report 0
```

Even if you have two different application servers running two different instances of your Python application, this transaction will fail, as desired, for the second customer, regardless of the point at which the second customer's transaction begins.

Two alternatives to stored procedures

As I noted earlier, I am not a fan of stored procedures. They have the advantage of executing faster that the same logic in an application language. Furthermore, multiserver transaction management through stored procedures is very elegant. But I have three key objections:

- Stored procedures are not portable from one database to another.
- They require an extended understanding of database programming—something that may not be available to all development teams.
- They don't completely solve the problem of scaling transactions across application servers under all scenarios. You still need to write your applications to use them wisely, and the result may, in fact, make your application more complicated.

In addition to these core objections, I personally strongly prefer a very strict separation of presentation, business modeling, business logic, and data.

The last objection is subjective and perhaps a nasty personal quirk. The first two objections, however, are real problems. After all, how many of you reading this book have found yourselves stuck with Oracle applications that could very easily work in MySQL if it weren't for all the stored procedures? You are paying a huge Oracle tax just because you used stored procedures to build your applications!

The second objection is a bit more esoteric. If you have the luxury of a large development staff with a diverse skill set, you don't see this problem. If you are in a small company that needs each person to wear multiple hats, it helps to have an application architecture that requires little or no database programming expertise.

To keep your logic at the application server level while still maintaining multiserver transactional integrity, you must either create protections against dirty writes or create a lock in the database.

In Chapter 3 of O'Reilly's *Java Database Best Practices (http://oreilly.com/catalog/9780596005221/index.html)*, I describe in detail transaction management policies for Java systems managing their transaction logic in the application server tier. One of the techniques I feature—and generally recommend for its speed benefits, whatever your overall architecture—is the use of a last update timestamp and modifying agent in your updates.

The booking logic from the stored procedure essentially was an update to the `booking` table:

```
UPDATE booking SET customer = ? WHERE booking_id = ?;
```

If you add last_update_timestamp and last_update_user fields, that SQL would operate more effectively in a multiserver environment:

```
UPDATE booking
SET customer = ?, last_update_timestamp = ?, last_update_user = ?
WHERE booking_id = ? AND last_update_timestamp = ? AND last_update_user = ?;
```

In this situation, the first client will attempt to book the room for the specified date and succeed. The second client then attempts to update the row but gets no matches since the timestamp it reads—as well as the user ID of the user on the client—will not match the values updated by the first client. The second client realizes it has updated zero rows and subsequently displays an error message. No double booking!

This approach works well as long as you do not end up structuring transactions in a way that will create *deadlocks*. A deadlock occurs between two transactions when each transaction is waiting on the other to release a lock. Our reservations system example is an application in which a deadlock is certainly possible.

Because we are booking a range of dates in the same transaction, poorly structured application logic could cause two clients to wait on each other as one attempts to book a date already booked by the other, and vice versa. For example, if you and I are looking to book both Tuesday and Wednesday, but for whatever reason your client first tries Wednesday and my client first tries Tuesday, we will end up in a deadlock where I wait on you to commit your Wednesday booking and you wait on me to commit my Tuesday booking.

This somewhat contrived scenario is easy to address by making sure that you move sequentially through each day. Other application logic, however, may not have as obvious a solution.

Another alternative is to create a field for managing your locks. The room table, for example, might have two extra columns for booking purposes: locked_by and locked_timestamp. Before starting the transaction that books the rooms, update the room table and commit the update. Once your booking transaction completes, release the lock by nulling out those fields prior to committing that transaction.

Because this approach requires two different database transactions, you are no longer executing the booking as a single atomic transaction. Consequently, you risk leaving an open lock that prevents others from booking any rooms on any dates. You can eliminate this problem through two tricks:

- The room is considered unlocked not only when the fields are NULL, but also when the locked_timestamp has been held for a long period of time.
- When updating the lock at the end of your booking transaction, use the locked_by and locked_timestamp fields in the WHERE clause. Thus, if someone else steals a lock out from under you, you only end up rolling back your transaction.

Both of these approaches are admittedly more complex than taking advantage of stored procedures. Regardless of what approach you use, however, the important key for the cloud

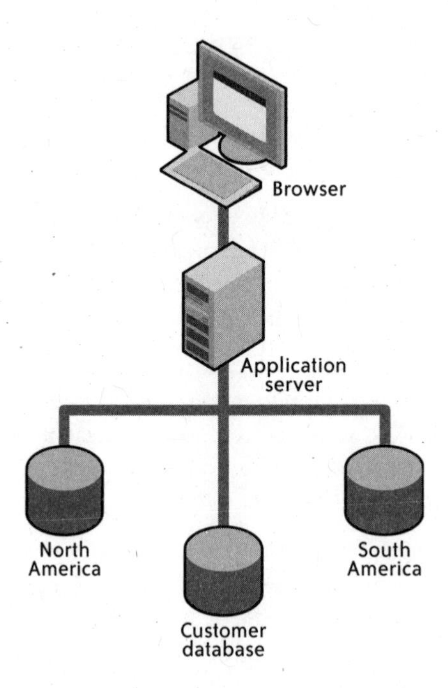

FIGURE 4-3. Supporting different hotels on different servers guarantees no double bookings

is simply making sure that you are not relying on memory locking to maintain your application state integrity.

When Servers Fail

The ultimate architectural objective for the cloud is to set up a running environment where the failure of any given application server ultimately doesn't matter. If you are running just one server, that failure will obviously matter at some level, but it will still matter less than losing a physical server.

One trick people sometimes use to get around the problems described in the previous section is data segmentation—also known as *sharding*. Figure 4-3 shows how you might use data segmentation to split processing across multiple application servers.

In other words, each application server manages a subset of data. As a result, there is never any risk that another server will overwrite the data. Although segmentation has its place in scaling applications, that place is not at the application server in a cloud cluster. A segmented application server cluster ultimately has a very low availability rating, as the failure of any individual server does matter.

The final corollary to all of this discussion of application state and server failure is that application servers in a cloud cannot store any state data beyond caching data. In other words, if you need to back up your application server, you have failed to create a solid application server architecture for the cloud. All state information—including binary data—belongs in the database, which must be on a persistent system.†

Machine Image Design

Two indirect benefits of the cloud are:

- It forces discipline in deployment planning
- It forces discipline in disaster recovery

Thanks to the way virtualized servers launch from machine images, your first step in moving into any cloud infrastructure is to create a repeatable deployment process that handles all the issues that could come up as the system starts up. To ensure that it does, you need to do some deployment planning.

The machine image (in Amazon, the AMI) is a raw copy of your operating system and core software for a particular environment on a specific platform. When you start a virtual server, it copies its operating environment from the machine image and boots up. If your machine image contains your installed application, deployment is nothing more than the process of starting up a new virtual instance.

Amazon Machine Image Data Security

When you create an Amazon machine image, it is encrypted and stored in an Amazon S3 bundle. One of two keys can subsequently decrypt the AMI:

- Your Amazon key
- A key that Amazon holds

Only your user credentials have access to the AMI. Amazon needs the ability to decrypt the AMI so it can actually boot an instance from the AMI.

DON'T STORE SENSITIVE DATA IN AN AMI

Even though your AMI is encrypted, I strongly recommend never storing any sensitive information in an AMI. Not only does Amazon have theoretical access to decrypt the AMI, but there also are

† If you have read my article "Ten MySQL Best Practices" (*http://www.onlamp.com/pub/a/onlamp/2002/07/11/MySQLtips.html*), you may have noted the contradiction between my admonition against storing binary data in MySQL and what I am recommending here. You should actually be caching the binary assets on the application server so that you do not need to pull them in real time from the database server. By doing so, you will get around my objections to storing binary data in MySQL.

mechanisms that enable you to make your AMI public and thus perhaps accidentally share whatever sensitive data you were maintaining in the AMI.

For example, if one company sues another Amazon customer, a court may subpoena the other Amazon customer's data. Unfortunately, it is not uncommon for courts to step outside the bounds of common sense and require a company such as Amazon to make available all Amazon customer data. If you want to make sure your data is never exposed as the result of a third-party subpoena, you should not store that data in an Amazon AMI.

Instead, encrypt it separately and load it into your instance at launch so that Amazon will not have the decryption keys and thus the data cannot be accessed, unless you are a party to the subpoena.

What Belongs in a Machine Image?

A machine image should include all of the software necessary for the runtime operation of a virtual instance based on that image *and nothing more*. The starting point is obviously the operating system, but the choice of components is absolutely critical. The full process of establishing a machine image consists of the following steps:

1. Create a component model that identifies what components and versions are required to run the service that the new machine image will support.
2. Separate out stateful data in the component model. You will need to keep it out of your machine image.
3. Identify the operating system on which you will deploy.
4. Search for an existing, trusted baseline public machine image for that operating system.
5. Harden your system using a tool such as Bastille.
6. Install all of the components in your component model.
7. Verify the functioning of a virtual instance using the machine image.
8. Build and save the machine image.

The starting point is to know exactly what components are necessary to run your service. Figure 4-4 shows a sample model describing the runtime components for a MySQL database server.

In this case, the stateful data exists in the MySQL directory, which is externally mounted as a block storage device. Consequently, you will need to make sure that your startup scripts mount your block storage device before starting MySQL.

Because the stateful data is assumed to be on a block storage device, this machine image is useful in starting *any* MySQL databases, not just a specific set of MySQL databases.

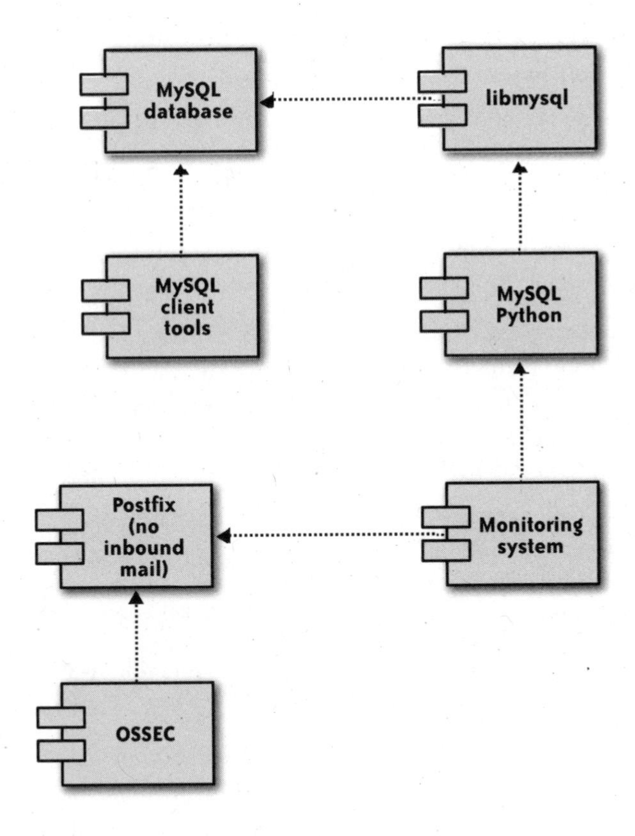

FIGURE 4-4. Software necessary to support a MySQL database server

The services you want to run on an instance generally dictate the operating system on which you will base the machine image. If you are deploying a .NET application, you probably will use one of the Amazon Windows images. A PHP application, on the other hand, probably will be targeting a Linux environment. Either way, I recommend searching some of the more trusted prebuilt, basic AMIs for your operating system of choice and customizing from there. Chapter 2 has a much deeper discussion of the technical details of creating an AMI.

> **WARNING**
>
> **Avoid using "kitchen sink" Linux distributions. Each machine image should be a hardened operating system and have only the tools absolutely necessary to serve its function.**

Hardening an operating system is the act of minimizing attack vectors into a server. Among other things, hardening involves the following activities:

- Removing unnecessary services.
- Removing unnecessary accounts.

- Running all services as a role account (not root) when possible.
- Running all services in a restricted jail when possible.
- Verifying proper permissions for necessary system services.

The best way to harden your Linux system is to use a proven hardening tool such as Bastille. I go into more detail on securing and hardening your cloud environments in Chapter 5.

Now that you have a secure base from which to operate, it is time to actually install the software that this system will support. In the case of the current example, it's time to install MySQL.

When installing your server-specific services, you may have to alter the way you think about the deployment thanks to the need to keep stateful data out of the machine image. For a MySQL server, you would probably keep stateful data on a block device and mount it at system startup. A web server, on the other hand, might store stateful media assets out in a cloud storage system such as Amazon S3 and pull it over into the runtime instance on startup.

Different applications will definitely require different approaches based on their unique requirements. Whatever the situation, you should structure your deployment so that the machine image has the intelligence to look for its stateful data upon startup and provide your machine image components with access to that data before they need it.

Once you have the deployment structured the right way, you will need to test it. That means testing the system from launch through shutdown and recovery. Therefore, you need to take the following steps:

1. Build a temporary image from your development instance.
2. Launch a new instance from the temporary image.
3. Verify that it functions as intended.
4. Fix any issues.
5. Repeat until the process is robust and reliable.

At some point, you will end up with a functioning instance from a well-structured machine image. You can then build a final instance and go have a beer (or coffee).

A Sample MySQL Machine Image

The trick to creating a machine image that supports database servers is knowing how your database engine of choice stores its data. In the case of MySQL, the database engine has a data directory for its stateful data. This data directory may actually be called any number of things (*/usr/local/mysql/data*, */var/lib/mysql*, etc.), but it is the only thing other than the configuration file that must be separated from your machine image. In a typical custom build, the data directory is */usr/local/mysql/data*.

NOTE

If you are going to be supporting multiple machine images, it often helps to first build a hardened machine image with no services and then build each service-oriented image from that base.

Once you start an instance from a standard image and harden it, you need to create an elastic block storage volume and mount it.‡ The standard Amazon approach is to mount the volume off of */mnt* (e.g., */mnt/database*). Where you mount it is technically unimportant, but it can help reduce confusion to keep the same directory for each image.

You can then install MySQL, making sure to install it within the instance's root filesystem (e.g., */usr/local/mysql*). At that point, move the data over into the block device using the following steps:

1. Stop MySQL if the installation process automatically started it.
2. Move your data directory over into your mount and give it a name more suited to mounting on a separate device (e.g., */mnt/database/mysql*).
3. Change your *my.cnf* file to point to the new data directory.

You now have a curious challenge on your hands: MySQL cannot start up until the block device has been mounted, but a block device under Amazon EC2 cannot be attached to an instance of a virtual machine until that instance is running. As a result, you cannot start MySQL through the normal boot-up procedures. However, you can end up where you want by enforcing the necessary order of events: boot the virtual machine, mount the device, and finally start MySQL. You should therefore carefully alter your MySQL startup scripts so that the system will no longer start MySQL on startup, but will still shut the MySQL engine down on shutdown.

WARNING

Do not simply axe MySQL from your startup scripts. Doing so will prevent MySQL from cleanly shutting down when you shut down your server instance. You will thus end up with a corrupt database on your block storage device.

The best way to effect this change is to edit the MySQL startup script to wait for the presence of the MySQL data directory before starting the MySQL executable.

Amazon AMI Philosophies

In approaching AMI design, you can follow one of two core philosophies:

- A minimalist approach in which you build a few multipurpose machine images.

‡ I describe this process for Amazon EC2 in Chapter 2.

- A comprehensive approach in which you build numerous purpose-specific machine images.

I am a strong believer in the minimalist approach. The minimalist approach has the advantage of being easier for rolling out security patches and other operating-system-level changes. On the flip side, it takes a lot more planning and EC2 skills to structure a multipurpose AMI capable of determining its function after startup and self-configuring to support that function. If you are just getting started with EC2, it is probably best to take the comprehensive approach and use cloud management tools to eventually help you evolve into a library of minimalist machine images.

For a single application installation, you won't likely need many machine images, and thus the difference between a comprehensive approach and a minimalist approach is negligible. SaaS applications—especially ones that are not multitenant—require a runtime deployment of application software.

Runtime deployment means uploading the application software—such as the MySQL executable discussed in the previous section—to a newly started virtual instance after it has started, instead of embedding it in the machine image. A runtime application deployment is more complex (and hence the need for cloud management tools) than simply including the application in the machine image, but it does have a number of major advantages:

- You can deploy and remove applications from a virtual instance while it is running. As a result, in a multiapplication environment, you can easily move an application from one cluster to another.
- You end up with automated application restoration. The application is generally deployed at runtime using the latest backup image. When you embed the application in an image, on the other hand, your application launch is only as good as the most recent image build.
- You can avoid storing service-to-service authentication credentials in your machine image and instead move them into the encrypted backup from which the application is deployed.

Privacy Design

In Chapter 5, I talk about all aspects of security and the cloud. As we design your overall application architecture in this chapter, however, it is important to consider how you approach an application architecture for systems that have a special segment of private data, notably e-commerce systems that store credit cards and health care systems with health data. We take a brief look at privacy design here, knowing that a full chapter of security awaits us later.

Privacy in the Cloud

The key to privacy in the cloud—or any other environment—is the strict separation of sensitive data from nonsensitive data followed by the encryption of sensitive elements. The simplest

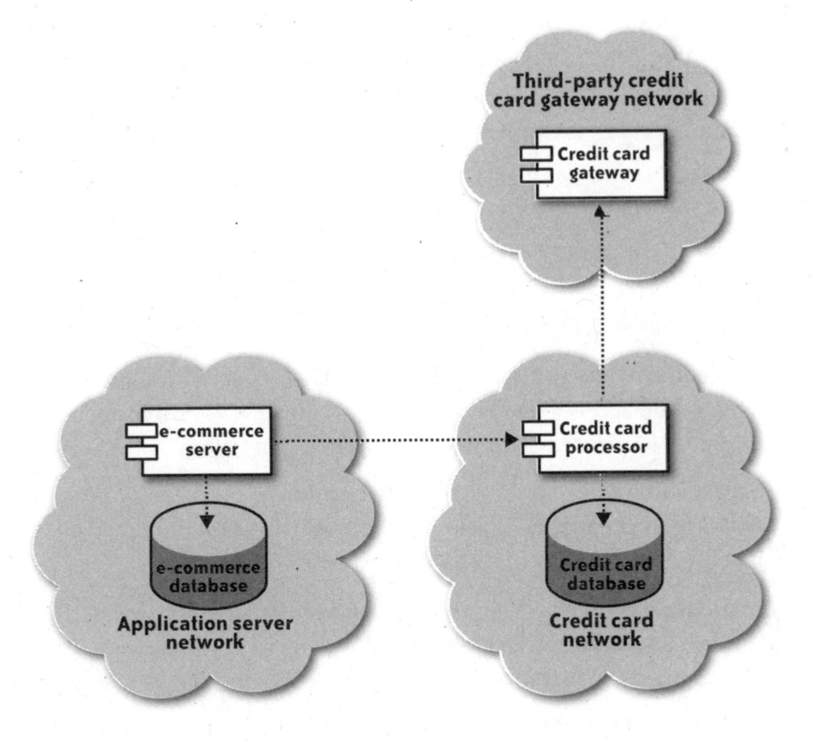

FIGURE 4-5. Host credit card data behind a web service that encrypts credit card data

example is storing credit cards. You may have a complex e-commerce application storing many data relationships, but you need to separate out the credit card data from the rest of it to start building a secure e-commerce infrastructure.

> **NOTE**
>
> When I say you need to separate the data, what I mean is that access to either of the two pieces of your data cannot compromise the privacy of the data. In the case of a credit card, you need to store the credit card number on a different virtual server in a different network segment and encrypt that number. Access to the first set of data provides only customer contact info; access to the credit card number provides only an encrypted credit card number.

Figure 4-5 provides an application architecture in which credit card data can be securely managed.

It's a pretty simple design that is very hard to compromise as long as you take the following precautions:

- The application server and credit card server sit in two different security zones with only web services traffic from the application server being allowed into the credit card processor zone.
- Credit card numbers are encrypted using a customer-specific encryption key.
- The credit card processor has no access to the encryption key, except for a short period of time (in memory) while it is processing a transaction on that card.
- The application server never has the ability to read the credit card number from the credit card server.
- No person has administrative access to both servers.

Under this architecture, a hacker has no use for the data on any individual server; he must hack both servers to gain access to credit card data. Of course, if your web application is poorly written, no amount of structure will protect you against that failing.

You therefore need to minimize the ability of a hacker to use one server to compromise the other. Because this problem applies to general cloud security, I cover it in detail in Chapter 5. For now, I'll just list a couple rules of thumb:

- Make sure the two servers have different attack vectors. In other words, they should not be running the same software. By following this guideline, you guarantee that whatever exploit compromised the first server is not available to compromise the second server.
- Make sure that neither server contains credentials or other information that will make it possible to compromise the other server. In other words, don't use passwords for user logins and don't store any private SSH keys on either server.

Managing the credit card encryption

In order to charge a credit card, you must provide the credit card number, an expiration date, and a varying number of other data elements describing the owner of the credit card. You may also be required to provide a security code.

This architecture separates the basic capture of data from the actual charging of the credit card. When a person first enters her information, the system stores contact info and some basic credit card profile information with the e-commerce application and sends the credit card number over to the credit card processor for encryption and storage.

The first trick is to create a password on the e-commerce server and store it with the customer record. It's not a password that any user will ever see or use, so you should generate something complex using the strongest password guidelines. You should also create a credit card record on the e-commerce server that stores everything except the credit card number. Figure 4-6 shows a sample e-commerce data model.

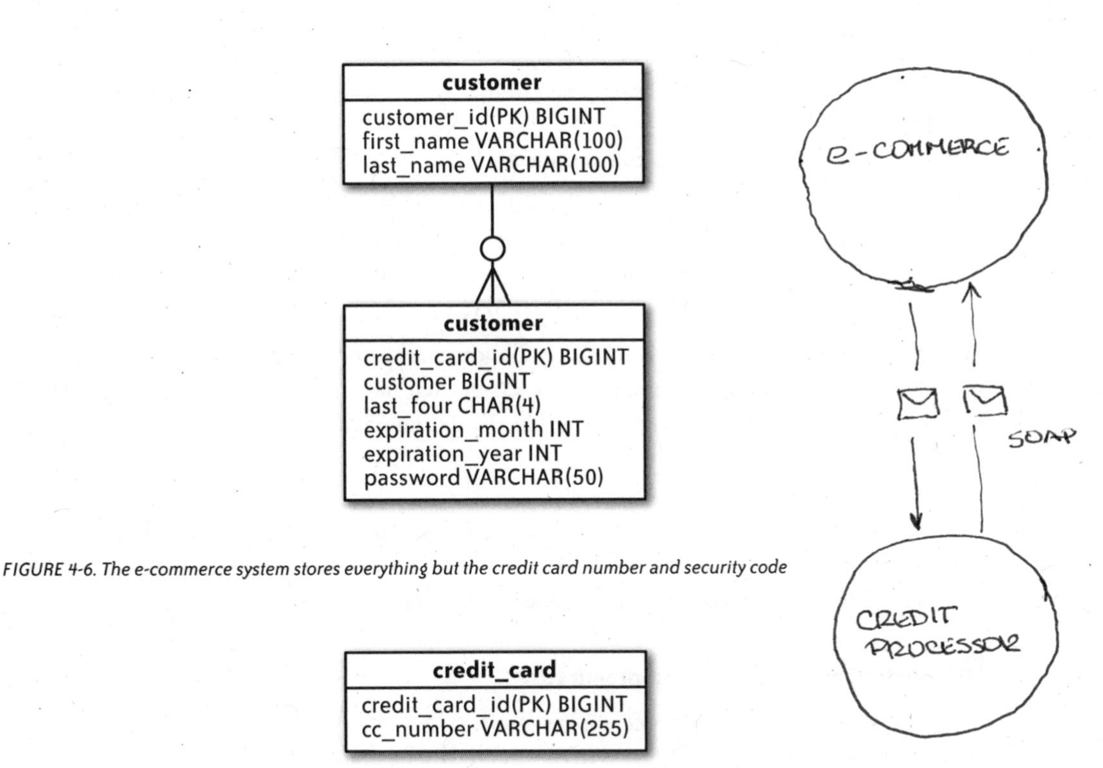

FIGURE 4-6. The e-commerce system stores everything but the credit card number and security code

FIGURE 4-7. The credit card processor stores the encrypted credit card number and associates it with the e-commerce credit card ID

With that data stored in the e-commerce system database, the system then submits the credit card number, credit card password, and unique credit card ID from the e-commerce system to the credit card processor.

The credit card processor does not store the password. Instead, it uses the password as salt to encrypt the credit card number, stores the encrypted credit card number, and associates it with the credit card ID. Figure 4-7 shows the credit card processor data model.

Neither system stores a customer's security code, because the credit card companies do not allow you to store this code.

Processing a credit card transaction

When it comes time to charge the credit card, the e-commerce service submits a request to the credit card processor to charge the card for a specific amount. The e-commerce system refers to the credit card on the credit card processor using the unique ID that was created when the credit card was first inserted. It passes over the credit card password, the security code, and the amount to be charged. The credit card processor then decrypts the credit card number for the specified credit card using the specified password. The unencrypted credit card number, security code, and amount are then passed to the bank to complete the transaction.

If the e-commerce application is compromised

If the e-commerce application is compromised, the attacker has access only to the nonsensitive customer contact info. There is no mechanism by which he can download that database and access credit card information or otherwise engage in identity theft. That would require compromising the credit card processor separately.

Having said all of that, if your e-commerce application is insecure, an attacker can still assume the identity of an existing user and place orders in their name with deliveries to their address. In other words, you still need to worry about the design of each component of the system.

> **NOTE**
>
> **Obviously, you don't want intruders gaining access to your customer contact data either. In the context of this section, my references to customer contact data as "nonsensitive" is a relative term. Your objective should be to keep an intruder from getting to either bit of data.**

If the credit card processor is compromised

Compromising the credit card processor is even less useful than compromising the e-commerce application. If an attacker gains access to the credit card database, all he has are random unique IDs and strongly encrypted credit card numbers—each encrypted with a unique encryption key. As a result, the attacker can take the database offline and attempt to brute-force decrypt the numbers, but each number will take a lot of time to crack and, ultimately, provide the hacker with a credit card number that has no individually identifying information to use in identity theft.

Another attack vector would be to figure out how to stick a Trojan application on the compromised server and listen for decryption passwords. However, if you are running intrusion detection software as suggested in Chapter 5, even this attack vector becomes unmanageable.

When the Amazon Cloud Fails to Meet Your Needs

The architecture I described in the previous section matches traditional noncloud deployments fairly closely. You may run into challenges deploying in the Amazon cloud, however, because of a couple of critical issues involving the processing of sensitive data:

- Some laws and specifications impose conditions on the political and legal jurisdictions where the data is stored. In particular, companies doing business in the EU may not store private data about EU citizens on servers in the U.S. (or any other nation falling short of EU privacy standards).

- Some laws and specifications were not written with virtualization in mind. In other words, they specify physical servers in cases where virtual servers would do identically well, simply because a server meant a physical server at the time the law or standard was written.

The first problem has a pretty clear solution: if you are doing business in the EU and managing private data on EU citizens, that data must be handled on servers with a physical presence in the EU, stored on storage devices physically in the EU, and not pass through infrastructure managed outside the EU.

Amazon provides a presence in both the U.S. and EU. As a result, you can solve the first problem by carefully architecting your Amazon solution. It requires, however, that you associate the provisioning of instances and storage of data with your data management requirements.

The second issue is especially problematic for solutions such as Amazon that rely entirely on virtualization. In this case, however, it's for fairly stupid reasons. You can live up to the spirit of the law or specification, but because the concept of virtualization was not common at the time, you cannot live up to the letter of the law or specification. The workaround for this scenario is similar to the workaround for the first problem.

In solving these challenges, you want to do everything to realize as many of the benefits of the cloud as possible without running private data through the cloud and without making the overall complexity of the system so high that it just isn't worth it. Cloud providers such as Rackspace and GoGrid tend to make such solutions easier than attempting a hybrid solution with Amazon and something else.

WHAT IS SENSITIVE?

When it comes to laws and standards, what you may think of as sensitive may not be what the law has in mind. In some cases, for example, EU privacy laws consider IP addresses to be personally identifying information. I am not a legal expert and therefore am not even going to make any attempts to define what information is private. But once you and your legal advisers have determined what must be protected as private, follow the general rule in this section: private data goes into the privacy-protected server and nonprivate data can go into the cloud.

To meet this challenge, you must route and store all private information outside the cloud, but execute as much application logic as possible inside the cloud. You can accomplish this goal by following the general approach I described for credit card processing and abstracting the concepts out into a privacy server and a web application server:

- The privacy server sits outside the cloud and has the minimal support structures necessary to handle your private data.

- The web application server sits inside the cloud and holds the bulk of your application logic.

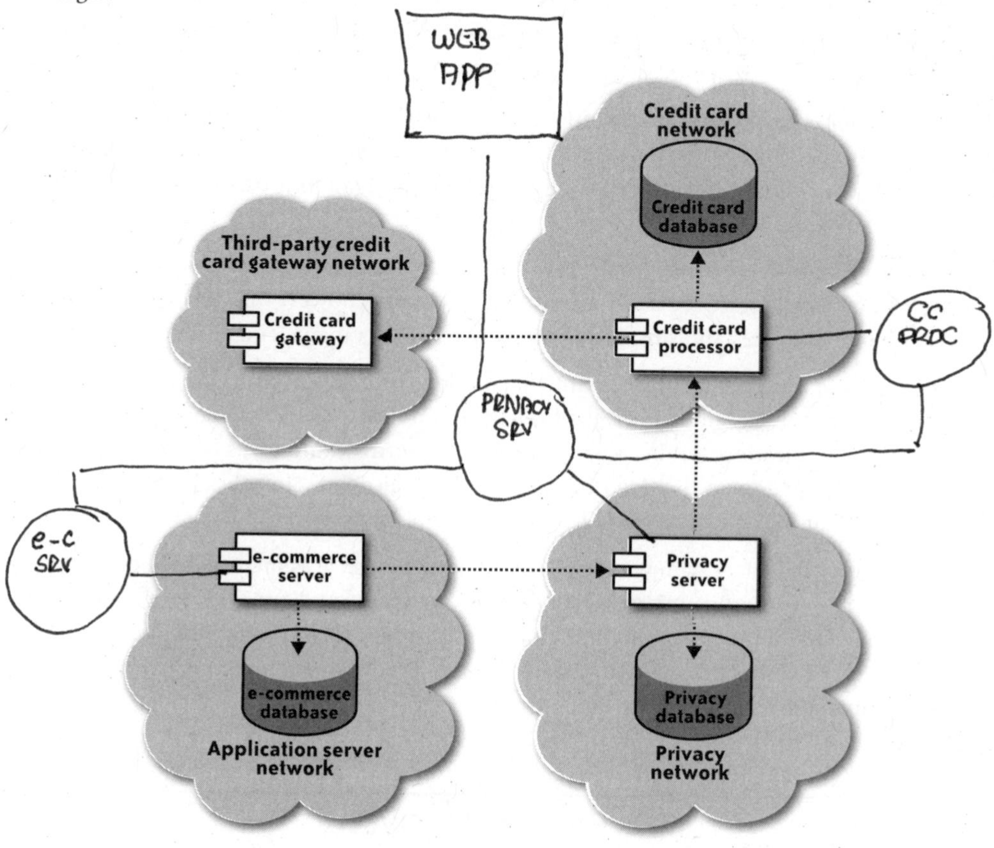

FIGURE 4-8. Pulling private data out of the cloud creates three different application components

Because the objective of a privacy server is simply to physically segment out private data, you do not necessarily need to encrypt everything on the privacy server. Figure 4-8 illustrates how the e-commerce system might evolve into a privacy architecture designed to store all private data outside of the cloud.

As with the cloud-based e-commerce system, you store credit card data on its own server in its own network segment. The only difference for the credit card processor is that this time it is outside of the cloud.

The new piece to this puzzle is the customer's personally identifying information. This data now exists on its own server outside of the cloud, but still separate from credit card data. When saving user profile information, those actions execute against the privacy server instead of the main web application. Under no circumstances does the main web application have any access to personally identifying information, unless that data is aggregated before being presented to the web application.

How useful this architecture is depends heavily on how much processing you are doing that has nothing to do with private data. If all of your transactions involve the reading and writing of private data, you gain nothing by adding this complexity. On the other hand, if the management of private data is just a tiny piece of the application, you can gain all of the advantages of the cloud for the other parts of the application while still respecting any requirements around physical data location.

Database Management

The trickiest part of managing a cloud infrastructure is the management of your persistent data. Persistent data is essentially any data that needs to survive the destruction of your cloud environment. Because you can easily reconstruct your operating system, software, and simple configuration files, they do not qualify as persistent data. Only the data that cannot be reconstituted qualify. If you are following my recommendations, this data lives in your database engine.

The problem of maintaining database consistency is not unique to the cloud. The cloud simply brings a new challenge to an old problem of backing up your database, because your database server in the cloud will be much less reliable than your database server in a physical infrastructure. The virtual server running your database will fail completely and without warning. Count on it.

Whether physical or virtual, when a database server fails, there is the distinct possibility that the files that comprise the database state will get corrupted. The likelihood of that disaster depends on which database engine you are using, but it can happen with just about any engine out there.

Absent of corruption issues, dealing with a database server in the cloud is very simple. In fact, it is much easier to recover from the failure of a server in a virtualized environment than in the physical world: simply launch a new instance from your database machine image, mount the old block storage device, and you are up and running.

> **NOTE**
>
> **Use block storage devices for database storage. Block storage devices provide the best performance option (better than local storage) and make for more flexible database backup strategies.**

Clustering or Replication?

The most effective mechanism for avoiding corruption is leveraging the capabilities of a database engine that supports true *clustering*. In a clustered database environment, multiple database servers act together as a single logical database server. The mechanics of this process vary from database engine to database engine, but the result is that a transaction committed to the cluster will survive the failure of any one node and maintain full data consistency. In fact, clients of the database will never know that a node went down and will be able to continue operating.

Unfortunately, database clustering is very complicated and generally quite expensive.

- Unless you have a skilled DBA on hand, you should not even consider undertaking the deployment of a clustered database environment.
- A clustered database vendor often requires you to pay for the most expensive licenses to use the clustering capabilities in the database management system (DBMS). Even if you are using MySQL clustering, you will have to pay for five machine instances to effectively run that cluster.
- Clustering comes with significant performance problems. If you are trying to cluster across distinct physical infrastructures—in other words, across availability zones—you will pay a hefty network latency penalty.

WARNING

Although there are few challenges in clustering a database in the cloud, one is significant: the I/O challenges inherent in virtualized systems. In particular, write operations in a clustered system are very network intensive. As a result, heavy write applications will perform significantly worse in a clustered virtualized environment than in a standard data center.

The alternative to clustering is *replication*. A replication-based database infrastructure generally has a main server, referred to as the *database master*. Client applications execute write transactions against the database master. Successful transactions are then replicated to *database slaves*.

Replication has two key advantages over clustering:

- It is generally much simpler to implement.
- It does not require an excessive number of servers or expensive licenses.

WARNING

MySQL replication is *not* a viable solution for anyone who absolutely, positively cannot lose one byte of data as a result of a server failure. If you are in this kind of business, you probably can afford clustering.

Unfortunately, replication is not nearly as reliable as clustering. A database master can, in theory, fail after it has committed a transaction locally but before the database slave has received it. In that event, you would have a database slave that is missing data. In fact, when a database master is under a heavy load, the database slave can actually fall quite far behind the master. If the master is somehow corrupted, it can also begin replicating corrupted data.

Apart from reliability issues, a replicated environment does not failover as seamlessly as a clustered solution. When your database master fails, clients using that master for write transactions cannot function until the master is recovered. On the other hand, when a node in a cluster fails, the clients do not notice the failure because the cluster simply continues processing transactions.

Using database clustering in the cloud

The good news, in general, is that the cloud represents few specific challenges to database clustering. The bad news is that every single database engine has a different clustering mechanism (or even multiple approaches to clustering) and thus an in-depth coverage of cloud-based clustering is beyond the scope of this book. I can, however, provide a few guidelines:

- A few cluster architectures exist purely for performance and not for availability. Under these architectures, single points of failure may still exist. In fact, the complexity of clustering may introduce additional points of failure.
- Clusters designed for high availability are often slower at processing individual write transactions, but they can handle much higher loads than standalone databases. In particular, they can scale to meet your read volume requirements.
- Some solutions—such as MySQL—may require a large number of servers to operate effectively. Even if the licensing costs for such a configuration are negligible, the cloud costs will add up.
- The dynamic nature of IP address assignment within a cloud environment may add new challenges in terms of configuring clusters and their failover rules.

Using database replication in the cloud

For most nonmission-critical database applications, replication is a "good enough" solution that can save you a lot of money and potentially provide you with opportunities for performance optimization. In fact, a MySQL replication system in the cloud can provide you with a flawless backup and disaster recovery system as well as availability that can almost match that of a cluster. Because the use of replication in the cloud can have such a tremendous impact compared to replication in a traditional data center, we'll go into a bit more detail on using replication in the cloud than we did with clustering.

Figure 4-9 shows a simple replication environment.

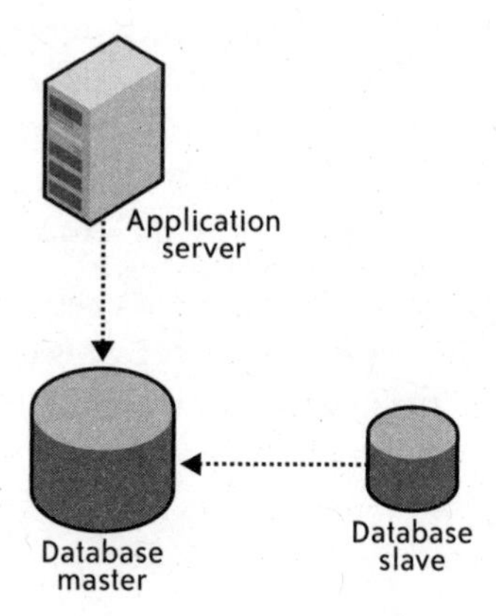

FIGURE 4-9. A simple replication (arrows show dependency)

In this structure, you have a single database server of record (the master) replicating to one or more copies (the slaves). In general,[§] the process that performs the replication from the master to the slave is not atomic with respect to the original transaction. In other words, just because a transaction successfully commits on the master does not mean that it successfully replicated to any slaves. The transactions that do make it to the slaves are generally atomic, so although a slave may be out of sync, the database on the slave should always be in an internally consistent state (uncorrupted).

Under a simple setup, your web applications point to the database master. Consequently, your database slave can fail without impacting the web application. To recover, start up a new database slave and point it to the master.

Recovering from the failure of a database master is much more complicated. If your cloud provider is Amazon, it also comes with some extra hurdles you won't see in a standard replication setup.

Ideally, you will recover your database master by starting a new virtual server that uses your database machine image and then mounting the volume that was formerly mounted by the failed server. The failure of your master, however, may have resulted in the corruption of the files on that volume. At this point, you will turn to the database slave.

A database can recover using a slave in one of two ways:

- Promotion of a slave to database master (you will need to launch a replacement slave)
- Building a new database master and exporting the current state from a slave to a new master

[§] This situation may not be the case for every configuration of every database engine in a replication architecture. Unless you know your replication is atomic with respect to the original transaction, however, assume your slave can be out of sync with the master at times.

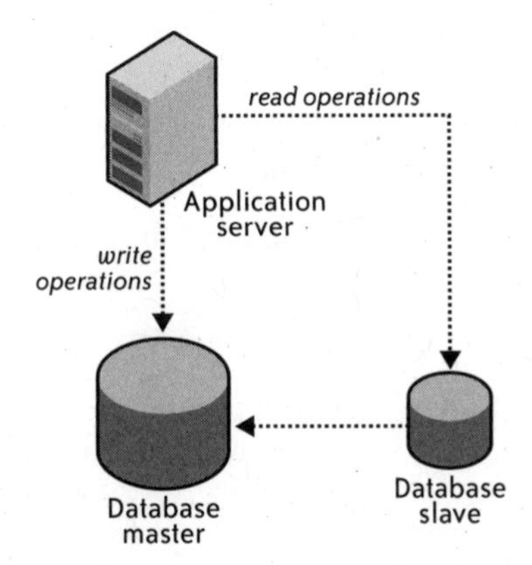

FIGURE 4-10. By separating read operations to execute against slaves, your applications can scale without clustering

Promotion is the fastest mechanism for recovery and the approach you almost certainly want to take, unless you have a need for managing distinct database master and database slave machine images. If that's the case, you may need to take the more complex recovery approach.

> **WARNING**
>
> To develop a bulletproof replication architecture, you need to look beyond recovery from the slave. It is possible that your slave process stopped running long before the master failed or, worse, that the slave went down with the master. You should therefore have the capability of restoring from a volume snapshot and, in the worst-case scenario, from a database dump in your cloud storage system. You should also have monitoring slave status as a key part of your cloud monitoring infrastructure.

As with other components in your web application architecture, putting your database in a replication architecture gives it the ability to rapidly recover from a node failure and, as a result, significantly increases overall system availability rating.

For a more detailed discussion of MySQL replication, I recommend reading O'Reilly's *High Performance MySQL (http://oreilly.com/catalog/9780596101718/index.html)*, by Baron Schwartz et al.

Replication for performance

Another reason to leverage replication is performance. Without segmenting your data, most database engines allow you to write against only the master, but you can read from the master or any of the slaves. An application heavy on read operations can therefore see significant performance benefits from spreading reads across slaves. Figure 4-10 illustrates the design of an application using replication for performance benefits.

The rewards of using replication for performance are huge, but there are also risks. The primary risk is that you might accidentally execute a write operation against one of the slaves. When you do that, replication falls apart and your master and slaves end up in inconsistent states. Two approaches to solving this problem include:

- Clearly separating read logic from write logic in your code and centralizing the acquisition of database connections.
- Making your slave nodes read-only.

The second one is the most foolproof, but it complicates the process of promoting a slave to master because you must reconfigure the server out of read-only mode before promoting it.

Primary Key Management

With a web application operating behind a load balancer in which individual nodes within the web application do not share state information with each other, the problem of cross-database primary key generation becomes a challenge. The database engine's auto-increment functionality is specific to the database you are using and not very flexible; it often is guaranteed to be unique only for a single server.

In *Java Database Best Practices* (O'Reilly; *http://oreilly.com/catalog/9780596005221/index.html*), I describe in detail a mechanism for generating keys in memory in an application server that are guaranteed to be unique across any number of application server nodes—even across multiple applications written in different languages. I'll cover that technique at a high level here and add a new twist: the generation of random identifiers that are guaranteed to be unique across multiple nodes.

How to generate globally unique primary keys

First, you could use standard UUIDs to serve as your primary key mechanism. They have the benefit of an almost nonexistent chance of generating conflicts, and most programming languages have built-in functions for generating them. I don't use them, however, for three reasons:

- They are 128-bit values and thus take more space and have longer lookup times than the 64-bit primary keys I prefer.
- Cleanly representing a 128-bit value in Java and some other programming languages is painful. In fact, the best way to represent such a value is through two separate values representing the 64 high bits and the 64 low bits, respectively.
- The possibility of collisions, although not realistic, does exist.

In order to generate identifiers at the application server level that are guaranteed to be unique in the target database, traditionally I rely on the database to manage key generation. I accomplish this through the creation of a `sequencer` table that hands out a key with a safe key space. The application server is then free to generate keys in that key space until the key space is exhausted.

> **NOTE**
>
> **I prefer to use 64-bit integers for primary keys in databases. 64 bits provide a large key space with fast lookups. The technique I talk about here works with alphanumeric key generation as well.**

The sequencer table looks like this:

```
CREATE TABLE sequencer (
    name          VARCHAR(20)        NOT NULL,
    next_key      BIGINT UNSIGNED    NOT NULL,
    last_update   BIGINT UNSIGNED    NOT NULL,
    spacing       INT    UNSIGNED    NOT NULL;
    PRIMARY KEY ( name, last_update ),
    UNIQUE INDEX ( name )
);
```

The first thing of note here is that there is nothing specific to any database in this table structure and your keys are not tied to a particular table. If necessary, multiple tables can share the same primary key space. Similarly, you can generate unique identifiers that have nothing to do with a particular table in your database.

To generate a unique `person_id` for your `person` table:

1. Set up a `next_key` value in memory and initialize it to `0`.
2. Grab the next `spacing` and `last_update` for the `sequencer` record with the `name = 'person.person_id'`.
3. Add `1` to the retrieved `next_key` and update the `sequencer` table with the `name` and retrieved `last_update` value in the `WHERE` clause.
4. If no rows are updated (because another server beat you to the punch), repeat steps 2 and 3.
5. Set the next person ID to `next_key`.
6. Increment the `next_key` value by 1.
7. The next time you need a unique person ID, simply execute steps 5 and 6 as long as `next_key < next_key + spacing`. Otherwise, set `next_key` to `0` and repeat the entire process.

Within the application server, this entire process must be locked against multithreaded access.

Support for globally unique random keys

The technique for unique key generation just described generates (more or less) sequential identifiers. In some cases, it is important to remove reasonable predictability from identifier generation. You therefore need to introduce some level of randomness into the equation.

WARNING

This technique is not truly random, but pseudorandom. There is no source of entropy in use for the randomly generated values, and the range of random values is sufficiently small for a determined guesser to break.

To get a random identifier, you need to multiply your next_key value by some power of 10 and then add a random number generated through the random number generator of your language of choice. The larger the random number possibility, the smaller your overall key space is likely to be. On the other hand, the smaller the random number possibility, the easier your keys will be to guess.

The following Python example illustrates how to generate a pseudorandom unique person ID:

```
import thread
import random

nextKey = -1;
spacing = 100;
lock = thread.allocate_lock();

def next():
    try:
        lock.acquire(); # make sure only one thread at a time can access
        if nextKey == -1 or nextKey > spacing:
            loadKey();
        nextId = (nextKey * 100000);
        nextKey = nextKey + 1;
    finally:
        lock.release();
    rnd = random.randint(0,99999);
    nextId = nextId + rnd;
    return nextId;
```

You can minimize the wasting of key space by tracking the allocation of random numbers and incrementing the nextKey value only after the random space has been sufficiently exhausted. The further down that road you go, however, the more likely you are to encounter the following challenges:

- The generation of unique keys will take longer.
- Your application will take up more memory.
- The randomness of your ID generation is reduced.

Database Backups

Throughout this book, I have hinted at the challenge of database backup management and its relationship to disaster recovery. I discuss the complete disaster recovery picture in the cloud in Chapter 6, but for now, I will deal with the specific problem of performing secure database backups in the cloud.

A good database backup strategy is hard, regardless of whether or not you are in the cloud. In the cloud, however, it is even more important to have a working database backup strategy.

Types of database backups

Most database engines provide multiple mechanisms for executing database backups. The rationale behind having different backup strategies is to provide a trade-off between the impact that executing a backup has on the production environment and the integrity of the data in the backup. Typically, your database engine will offer at least these backup options (in order of reliability):

- Database export/dump backup
- Filesystem backup
- Transaction log backup

> **NOTE**
>
> **Your database engine almost certainly provides other options. It may be valuable to tailor your backup processes to take advantage of those capabilities.**

The most solid backup you can execute is the database export/dump. When you perform a database export, you dump the entire schema of the database and all of its data to one or more export files. You can then store the export files as the backup. During recovery, you can leverage the export files to restore into a pristine install of your database engine.

To execute a database export on SQL Server, for example, use the following command:

```
BACKUP DATABASE website to disk = 'D:\db\website.dump'
```

The result is an export file you can move from one SQL Server environment to another SQL Server environment.

The downside of the database export is that your database server must be locked against writes in order to get a complete export that is guaranteed to be in an internally consistent state. Unfortunately, the export of a large database takes a long time to execute. As a result, full database exports against a production database generally are not practical.

Most databases provide the option to export parts of the database individually. For example, you could dump just your access_log table every night. In MySQL:

```
$ mysqldump website access_log > /backups/db/website.dump
```

If the table has any dependencies on other tables in the system, however, you can end up with inconsistent data when exporting on a table-by-table basis. Partial exports are therefore most useful on data from a data warehouse.

Filesystem backups involve backing up all of the underlying files that support the database. For some database engines, the database is stored in one big file. For others, the tables and their schemas are stored across multiple files. Either way, a backup simply requires copying the database files to backup media.

Though a filesystem backup requires you to lock the database against updates, the lock time is typically shorter. In fact, the snapshotting capabilities of block storage devices generally reduce the lock time to under a second, no matter how large the database is.

The following SQL will freeze MySQL and allow you to snapshot the filesystem on which the database is stored:

```
FLUSH TABLES WITH READ LOCK
```

With the database locked, take a snapshot of the volume, and then release the lock.

The least disruptive kind of backup is the transaction log backup. As a database commits transactions, it writes those transactions to a transaction logfile. Because the transaction log contains only committed transactions, you can back up these transaction logfiles without locking the database or stopping. They are also smaller files and thus back up quickly. Using this strategy, you will create a full database backup on a nightly or weekly basis and then back up the transaction logs on a more regular basis.

Restoring from transaction logs involves restoring from the most recent full database backup and then applying the transaction logs. This approach is a more complex backup scheme than the other two because you have a number of files created at different times that must be managed together. Furthermore, restoring from transaction logs is the longest of the three restore options.

Applying a backup strategy for the cloud

The best backup strategy for the cloud is a file-based backup solution. You lock the database against writes, take a snapshot, and unlock it. It is elegant, quick, and reliable. The key cloud feature that makes this approach possible is the cloud's ability to take snapshots of your block storage volumes. Without snapshot capabilities, this backup strategy would simply take too long.

Your backup strategy cannot, however, end with a file-based backup. Snapshots work beautifully within a single cloud, but they cannot be leveraged outside your cloud provider. In other words, an Amazon EC2 elastic block volume snapshot cannot be leveraged in a cloud deployment. To make sure your application is portable between clouds, you need to execute full database exports regularly.

How regularly you perform your database exports depends on how much data you can use. The underlying question you need to ask is, "If my cloud provider suddenly goes down for an extended period of time, how much data can I afford to lose when launching in a new environment?"

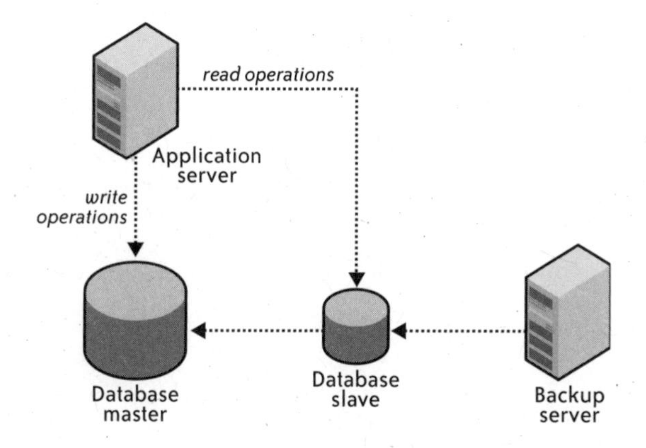

FIGURE 4-11. Execute regular full database exports against a replication slave

For a content management system, it may be OK in such an extreme situation to lose a week of data. An e-commerce application, however, cannot really afford to lose any data—even under such extreme circumstances.

My approach is to regularly execute full database exports against a MySQL slave, as shown in Figure 4-11.

For the purposes of a backup, it does not matter if your database slave is a little bit behind the master. What matters is that the slave represents the consistent state of the entire database at a relatively reasonable point in time. You can therefore execute a very long backup against the slave and not worry about the impact on the performance of your production environment. Because you can execute long backups, you can also execute numerous backups bounded mostly by your data storage appetite.

If your database backups truly take such a long time to execute that you risk having your slaves falling very far behind the master, it makes sense to configure multiple slaves and rotate backups among the slaves. This rotation policy will give a slave sufficient time to catch up with the master after it has executed a backup and before it needs to perform its next backup.

Once the backup is complete, you should move it over to S3 and regularly copy those backups out of S3 to another cloud provider or your own internal file server.

Your application architecture should now be well structured to operate not only in the Amazon cloud, but in other clouds as well.

CHAPTER FIVE

Security

IF THE CLOUD FORCES YOU TO COMPLETELY REEXAMINE YOUR THINKING about any particular part of your infrastructure, it's most likely to be security. The first question I hear from most executives is, "Should I be concerned about losing control over where my data is stored?" Although outsiders are particularly focused on this question, the following security implications of the cloud are much more profound:

- Lawsuits that do not involve you become a security concern.
- Many of the laws and standards that govern your IT infrastructure were created without virtualization in mind.
- The idea of perimeter security is largely nonsensical in the cloud.
- How you manage user credentials goes beyond standard identity management.

As with many other aspects of the cloud, security here can actually be better than in an internal data center. The ephemeral nature of virtual instances forces you to adopt robust security processes that many traditional hosting environments get away without using, so the move can result in a high-security computing infrastructure.

Data Security

Physical security defines how you control physical access to the servers that support your infrastructure. The cloud still has physical security constraints. After all, there are actual servers running somewhere. When selecting a cloud provider, you should understand their physical security protocols and the things you need to do on your end to secure your systems against physical vulnerabilities.

Data Control

The big chasm between traditional data centers and the cloud is the location of your data on someone else's servers. Companies who have outsourced their data centers to a managed services provider may have crossed part of that chasm; what cloud services add is the inability to see or touch the servers on which their data is hosted. The meaning of this change is a somewhat emotional matter, but it does present some real business challenges.

The main practical problem is that factors that have nothing to do with your business can compromise your operations and your data. For example, any of the following events could create trouble for your infrastructure:

- The cloud provider declares bankruptcy and its servers are seized or it ceases operations.
- A third party with no relationship to you (or, worse, a competitor) sues your cloud provider and obtains a blanket subpoena granting access to all servers owned by the cloud provider.
- Failure of your cloud provider to properly secure portions of its infrastructure—especially in the maintenance of physical access controls—results in the compromise of your systems.

The solution is to do two things you should be doing anyway, but likely are pretty lax about: encrypt everything and keep off-site backups.

- Encrypt sensitive data in your database and in memory. Decrypt it only in memory for the duration of the need for the data. Encrypt your backups and encrypt all network communications.
- Choose a second provider and use automated, regular backups (for which many open source and commercial solutions exist) to make sure any current and historical data can be recovered even if your cloud provider were to disappear from the face of the earth.

Let's examine how these measures deal with each scenario, one by one.

When the cloud provider goes down

This scenario has a number of variants: bankruptcy, deciding to take the business in another direction, or a widespread and extended outage. Whatever is going on, you risk losing access to your production systems due to the actions of another company. You also risk that the organization controlling your data might not protect it in accordance with the service levels to which they may have been previously committed.

I'll talk more in Chapter 6 about how to set up your backups and recover from this scenario. The bottom line, however, is that regular "off-site" backups will protect you here. Having a second cloud provider through which you can launch a replacement infrastructure is even better.

When a subpoena compels your cloud provider to turn over your data

If the subpoena is directed at you, obviously you have to turn over the data to the courts, regardless of what precautions you take, but these legal requirements apply whether your data is in the cloud or on your own internal IT infrastructure. What we're dealing with here is a subpoena aimed at your cloud provider that results from court action that has nothing to do with you.

Technically, a subpoena should be narrow enough that it does not involve you. You cannot, however, be sure that a subpoena relating to cutting-edge technology will be properly narrow, nor even that you'll know the subpoena has been issued.

Encrypting your data will protect you against this scenario. The subpoena will compel your cloud provider to turn over your data and any access it might have to that data, but your cloud provider won't have your access or decryption keys. To get at the data, the court will have to come to you and subpoena you. As a result, you will end up with the same level of control you have in your private data center.

When your cloud provider fails to adequately protect their network

When you select a cloud provider, you absolutely must understand how they treat physical, network, and host security. Though it may sound counterintuitive, the most secure cloud provider is one in which you never know where the physical server behind your virtual instance is running. Chances are that if you cannot figure it out, a determined hacker who is specifically targeting your organization is going to have a much harder time breaching the physical environment in which your data is hosted.

> **NOTE**
>
> Amazon does not even disclose where their data centers are located; they simply claim that each data center is housed in a nondescript building with a military-grade perimeter. Even if you know that my database server is in the `us-east-1a` availability zone, you don't know where the data center(s) behind that availability zone is located, or even which of the three East Coast availability zones `us-east-1a` represents.

Amazon publishes its security standards and processes at *http://aws.amazon.com*. Whatever cloud provider you use, you should understand their security standards and practices, and expect them to exceed anything you require.

Nothing guarantees that your cloud provider will, in fact, live up to the standards and processes they profess to support. If you follow everything else I recommend in this chapter, however, your data confidentiality will be strongly protected against even complete incompetence on the part of your cloud provider.

Encrypt Everything

In the cloud, your data is stored somewhere; you just don't know exactly where. However, you know some basic parameters:

- Your data lies within a virtual machine guest operating system, and you control the mechanisms for access to that data.
- Network traffic exchanging data between instances is not visible to other virtual hosts.
- For most cloud storage services, access to data is private by default. Many, including Amazon S3, nevertheless allow you to make that data public.

Encrypt your network traffic

No matter how lax your current security practices, you probably have network traffic encrypted—at least for the most part. A nice feature of the Amazon cloud is that virtual servers cannot sniff the traffic of other virtual servers. I still recommend against relying on this feature, since it may not be true of other providers. Furthermore, Amazon might roll out a future feature that renders this protection measure obsolete. You should therefore encrypt all network traffic, not just web traffic.

Encrypt your backups

When you bundle your data for backups, you should be encrypting it using some kind of strong cryptography, such as PGP. You can then safely store it in a moderately secure cloud storage environment like Amazon S3, or even in a completely insecure environment.

Encryption eats up CPU. As a result, I recommend first copying your files in plain text over to a temporary backup server whose job it is to perform encryption, and then uploading the backups into your cloud storage system. Not only does the use of a backup server avoid taxing your application server and database server CPUs, it also enables you to have a single higher-security system holding your cloud storage access credentials rather than giving those credentials to every system that needs to perform a backup.

Encrypt your filesystems

Each virtual server you manage will mount ephemeral storage devices (such as the */mnt* partition on Unix EC2 instances) or block storage devices. The failure to encrypt ephemeral devices poses only a very moderate risk in an EC2 environment because the EC2 Xen system zeros out that storage when your instance terminates. Snapshots for block storage devices, however, sit in Amazon S3 unencrypted unless you take special action to encrypt them.

NOTE

Encryption in all of its forms is expensive. Nowhere is the expense more of an issue than at the filesystem level. My default recommendation is to encrypt your filesystem, but that might not be practical for some applications. In the end, you must balance the performance requirements of your specific applications with its data protection requirements. Unfortunately, those requirements are likely in conflict at some level, and you may have to make a compromise to stay in the cloud.

The most secure approach to both scenarios is to mount ephemeral and block storage devices using an encrypted filesystem. Managing the startup of a virtual server using encrypted filesystems ultimately ends up being easier in the cloud and offers more security.

The challenge with encrypted filesystems on servers lies in how you manage the decryption password. A given server needs your decryption password before it can mount any given encrypted filesystem. The most common approach to this problem is to store the password on an unencrypted root filesystem. Because the objective of filesystem encryption is to protect against physical access to the disk image, the storage of the password on a separate, unencrypted filesystem is not as problematic as it might appear on the face of it—but it's still problematic.

In the cloud, you don't have to store the decryption password in the cloud. Instead, you can provide the decryption password to your new virtual instance when you start it up. The server can then grab the encryption key out of the server's startup parameters and subsequently mount any ephemeral or block devices using an encrypted filesystem.

You can add an extra layer of security into the mix by encrypting the password and storing the key for decrypting the password in the machine image. Figure 5-1 illustrates the process of starting up a virtual server that mounts an encrypted filesystem using an encrypted password.

Regulatory and Standards Compliance

Most problems with regulatory and standards compliance lie not with the cloud, but in the fact that the regulations and standards written for Internet applications predate the acceptance of virtualization technologies. In other words, chances are you can meet the spirit of any particular specification, but you may not be able to meet the letter of the specification.

For example, if your target standard requires certain data to be stored on a different server than other system logic, can a virtualized server ever meet that requirement? I would certainly argue that it *should* be able to meet that requirement, but the interpretation as to whether it does may be left up to lawyers, judges, or other nontechnologists who don't appreciate the nature of virtualization. It does not help that some regulations such as Sarbanes-Oxley (SOX) do not really provide any specific information security requirements, and seem to exist mostly for consultants to make a buck spreading fear among top-level management.

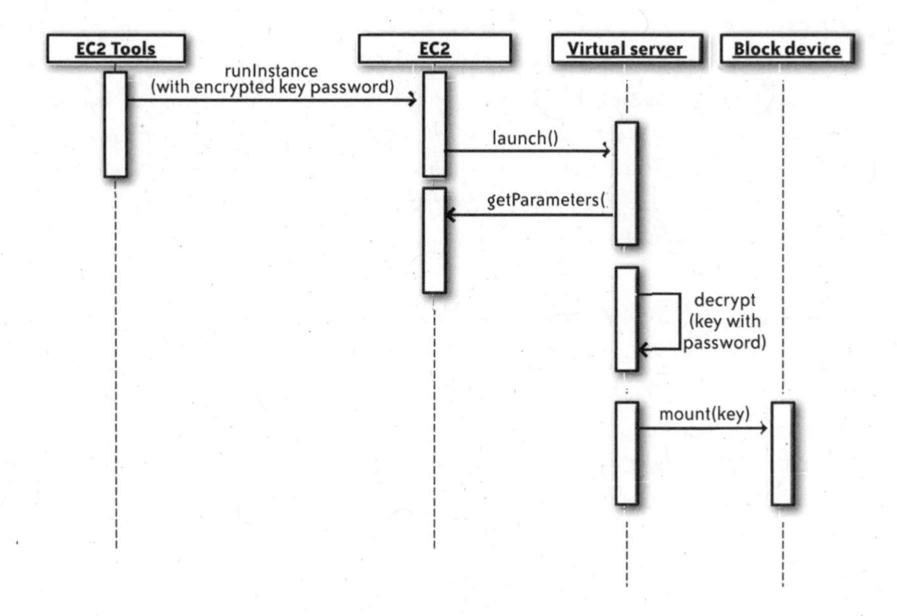

FIGURE 5-1. The process of starting a virtual server with encrypted filesystems

ALPHABET SOUP

Directive 95/46/EC
: EC Directive on Data Protection. A 1995 directive for European Union nations relating to the protection of private data and where it can be shared.

HIPAA
: Health Insurance Portability and Accountability Act. A comprehensive law relating to a number of health care issues. Of particular concern to technologists are the privacy and security regulations around the handling of health care data.

PCI or PCI DSS
: Payment Card Industry Data Security Standard. A standard that defines the information security processes and procedures to which an organization must adhere when handling credit card transactions.

SOX
: Sarbanes-Oxley Act. Establishes legal requirements around the reporting of publicly held companies to their shareholders.

21CFR11

Title 21 CFR Part 11 of the Federal Code of Regulations. This standard specifies guidelines governing electronic signatures and the management of electronic records in the pharmaceutical and medical device manufacturing industry.

From a security perspective, you'll encounter three kinds of issues in standards and regulations:

"How" issues

These result from a standard such as PCI or regulations such as HIPAA or SOX, which govern how an application of a specific type should operate in order to protect certain concerns specific to its problem domain. For example, HIPAA defines how you should handle personally identifying health care data.

"Where" issues

These result from a directive such as Directive 95/46/EC that governs where you can store certain information. One key impact of this particular directive is that the private data on EU citizens may not be stored in the United States (or any other country that does not treat private data in the same way as the EU).

"What" issues

These result from standards prescribing very specific components to your infrastructure. For example, PCI prescribes the use of antivirus software on all servers processing credit card data.

The bottom line today is that a cloud-deployed system may or may not be able to meet the letter of the law for any given specification. For certain specifications, you may be able to meet the letter of the specification by implementing a mixed architecture that includes some physical elements and some virtual elements. Cloud infrastructures that specialize in hybrid solutions may ultimately be a better solution. Alternatively, it may make sense to look at vendors who provide as a service the part of your system that has specific regulatory needs. For example, you can use an e-commerce vendor to handle the e-commerce part of your website and manage the PCI compliance issues.

In a mixed environment, you don't host any sensitive data in the cloud. Instead, you offload processing onto privacy servers in a physical data center in which the hosts are entirely under your control. For example, you might have a credit card processing server at your managed services provider accepting requests from the cloud to save credit card numbers or charge specific cards.

With respect to "where" data is stored, Amazon provides S3 storage in the EU. Through the Amazon cloud and S3 data storage, you do have the ability to achieve Directive 95/46/EC compliance with respect to storing data in the EU without building out a data center located in the EU.

Network Security

Amazon's cloud has no perimeter. Instead, EC2 provides security groups that define firewall-like traffic rules governing what traffic can reach virtual servers in that group. Although I often speak of security groups as if they were virtual network segments protected by a firewall, they most definitely are not virtual network segments, due to the following:

- Two servers in two different Amazon EC2 availability zones can operate in the same security group.
- A server may belong to more than one security group.
- Servers in the same security group may not be able to talk to each other at all.
- Servers in the same network segment may not share any IP characteristics—they may even be in different class address spaces.
- No server in EC2 can see the network traffic bound for other servers (this is not necessarily true for other cloud systems). If you try placing your virtual Linux server in promiscuous mode, the only network traffic you will see is traffic originating from or destined for your server.

Firewall Rules

Typically, a firewall protects the perimeter of one or more network segments. Figure 5-2 illustrates how a firewall protects the perimeter.

A main firewall protects the outermost perimeter, allowing in only HTTP, HTTPS, and (sometimes) FTP* traffic. Within that network segment are border systems, such as load balancers, that route traffic into a DMZ protected by another firewall. Finally, within the DMZ are application servers that make database and other requests across a third firewall into protected systems on a highly sensitive internal network.

This structure requires you to move through several layers—or perimeters—of network protection in the form of firewalls to gain access to increasingly sensitive data. The perimeter architecture's chief advantage is that a poorly structured firewall rule on the inner perimeter does not accidentally expose the internal network to the Internet unless the DMZ is already compromised. In addition, outer layer services tend to be more hardened against Internet

* Don't let FTP traffic in your network. FTP is an insecure protocol with a long history of vulnerabilities in the various FTP server implementations. Use SCP.

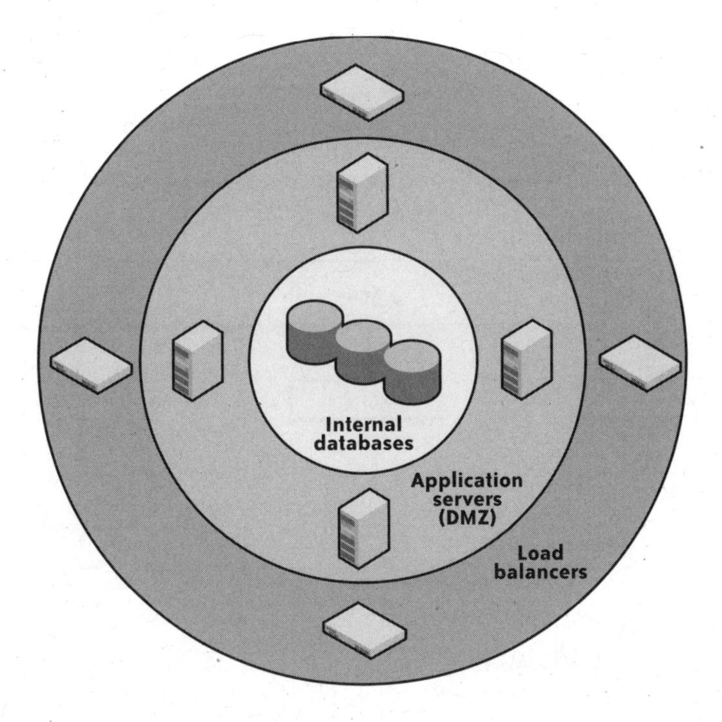

FIGURE 5-2. Firewalls are the primary tool in perimeter security

vulnerabilities, whereas interior services tend to be less Internet-aware. The weakness of this infrastructure is that a compromise of any individual server inside any given segment provides full access to all servers in that network segment.

Figure 5-3 provides a visual look at how the concept of a firewall rule in the Amazon cloud is different from that in a traditional data center.

Each virtual server occupies the same level in the network, with its traffic managed through a security group definition. There are no network segments, and there is no perimeter. Membership in the same group does not provide any privileged access to other servers in that security group, unless you define rules that provide privileged access. Finally, an individual server can be a member of multiple security groups. The rules for a given server are simply the union of the rules assigned to all groups of which the server is a member.

You can set up security groups to help you mimic traditional perimeter security. For example, you can create the following:

- A border security group that listens to all traffic on ports 80 and 443
- A DMZ security group that listens to traffic from the border group on ports 80 and 443
- An internal security group that listens to traffic on port 3306 from the DMZ security group

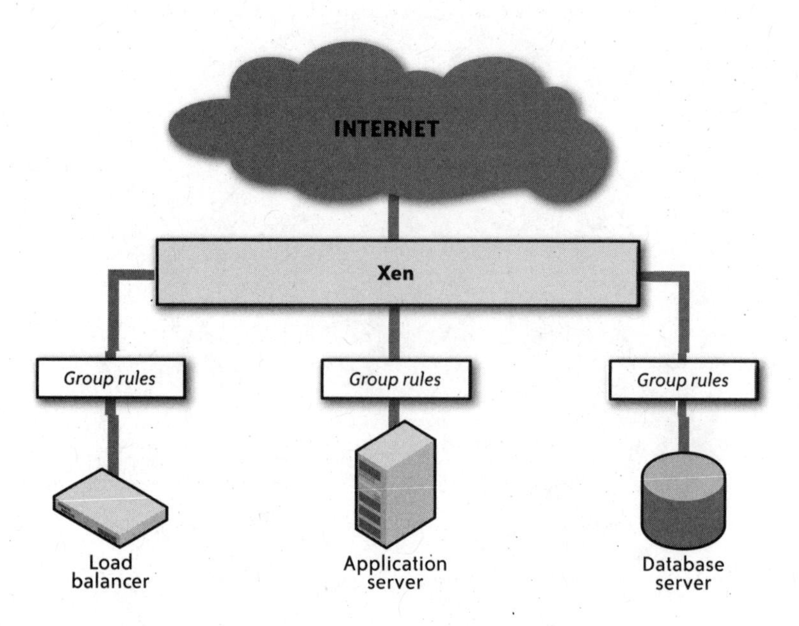

FIGURE 5-3. There are no network segments or perimeters in the cloud

WARNING

Amazon EC2 security does not currently enable you to limit access by port when defining group-to-group access. It may appear in the future. You can, however, mimic this behavior by defining source IP-based rules for each server in the source group.

As with traditional perimeter security, access to the servers in your internal security group requires first compromising the outer group, then the DMZ, and then finally one of the internal servers. Unlike traditional perimeter security, there is the possibility for you to accidentally grant global access into the internal zone and thus expose the zone. However, an intruder who compromises a single server within any given zone gains no ability to reach any other server in that zone except through leveraging the original exploit. In other words, access to the zone itself does not necessarily provide access to the other servers in that zone.

The Amazon approach also enables functionality that used to be out of the question in a traditional infrastructure. For example, you can more easily provide for direct SSH access into each virtual server in your cloud infrastructure from your corporate IT network without relying on a VPN. You still have the security advantages of a traditional perimeter approach when it comes to the open Internet, but you can get quick access to your servers to manage them from critical locations.

Two other advantages of this security architecture are the following:

- Because you control your firewall rules remotely, an intruder does not have a single target to attack, as he does with a physical firewall.
- You don't have the opportunity to accidentally destroy your network rules and thus permanently remove everyone's access to a given network segment.

I recommend the approach of mimicking traditional perimeter security because it is a well-understood approach to managing network traffic and it works. If you take that approach, it's important to understand that you are creating physical counterparts to the network segments of a traditional setup. You don't really have the layers of network security that come with a traditional configuration.

A few best practices for your network security include:

Run only one network service (plus necessary administrative services) on each virtual server
: Every network service on a system presents an attack vector. When you stick multiple services on a server, you create multiple attack vectors for accessing the data on that server or leveraging that server's network access rights.

Do not open up direct access to your most sensitive data
: If getting access to your customer database requires compromising a load balancer, an application server, and a database server (and you're running only one service per server), an attacker needs to exploit three different attack vectors before he can get to that data.

Open only the ports absolutely necessary to support a server's service and nothing more
: Of course your server should be hardened so it is running only the one service you intend to run on it. But sometimes you inadvertently end up with services running that you did not intend, or there is a nonroot exploit in the service you are running that enables an attacker to start up another service with a root exploit. By blocking access to everything except your intended service, you prevent these kinds of exploits.

Limit access to your services to clients who need to access them
: Your load balancers naturally need to open the web ports 80 and 443 to all traffic. Those two protocols and that particular server, however, are the only situations that require open access. For every other service, traffic should be limited to specific source addresses or security groups.

Even if you are not doing load balancing, use a reverse proxy
: A reverse proxy is a web server such as Apache that proxies traffic from a client to a server. By using a proxy server, you make it much harder to attack your infrastructure. First of all, Apache and IIS are much more battle-hardened than any of the application server options you will be using. As a result, an exploit is both less likely and almost certain to be patched more quickly. Second, an exploit of a proxy provides an attacker with access to nothing at all. They must subsequently find an additional vulnerability in your application server itself.

Use the dynamic nature of the cloud to automate your security embarrassments
: Admit it. You have opened up ports in your firewall to accomplish some critical business task even though you know better. Perhaps you opened an FTP port to a web server because some client absolutely had to use anonymous FTP for their batch file uploads. Instead of leaving that port open 24/7, you could open the port only for the batch window and then shut it down. You could even bring up a temporary server to act as the FTP server for the batch window, process the file, and then shut down the server.

The recommendations in the preceding list are not novel; they are standard security precautions. The cloud makes them relatively easy to implement, and they are important to your security there.

Network Intrusion Detection

Perimeter security often involves *network intrusion detection systems* (NIDS), such as Snort, which monitor local traffic for anything that looks irregular. Examples of irregular traffic include:

- Port scans
- Denial-of-service attacks
- Known vulnerability exploit attempts

You perform network intrusion detection either by routing all traffic through a system that analyzes it or by doing passive monitoring from one box on local traffic on your network. In the Amazon cloud, only the former is possible; the latter is meaningless since an EC2 instance can see only its own traffic.

The purpose of a network intrusion detection system

Network intrusion detection exists to alert you of attacks before they happen and, in some cases, foil attacks as they happen. Because of the way the Amazon cloud is set up, however, many of the things you look for in a NIDS are meaningless. For example, a NIDS typically alerts you to port scans as evidence of a precursor to a potential future attack. In the Amazon cloud, however, you are not likely to notice a port scan because your NIDS will be aware only of requests coming in on the ports allowed by your security group rules. All other traffic will be invisible to the NIDS and thus are not likely to be perceived as a port scan.

PORT SCANS AND THE AMAZON CLOUD

When an attacker is looking for vulnerabilities against a particular target, one of the first things they do is execute a port scan against a known server and then examine servers with nearby IP addresses. This approach does not provide terribly useful data when executed against the cloud for a number of reasons:

- Nodes with proximate IP addresses are almost always unrelated. As a result, you cannot learn anything about the network architecture of a particular organization by executing a port scan.
- Amazon security groups deny all incoming traffic by default, and requests for ports that have not been opened simply do not respond. As a result, very few ports for any particular server will actually be open. Furthermore, scanning across all ports is a very slow process because each closed port times out instead of actively denying the traffic.
- Amazon has its own intrusion detection systems in place and does not allow its customers to execute port scans against their own servers. As a result, an active port scan is likely to be blocked before any real information can be gathered.

As with port scans, Amazon network intrusion systems are actively looking for denial-of-service attacks and would likely identify any such attempts long before your own intrusion detection software.

One place in which an additional network intrusion detection system is useful is its ability to detect malicious payloads coming into your network. When the NIDS sees traffic that contains malicious payload, it can either block the traffic or send out an alert that enables you to react. Even if the payload is delivered and compromises a server, you should be able to respond quickly and contain the damage.

WARNING

Don't execute your own port scans against your cloud servers. Typically, when you harden your network infrastructure, you will use a tool such as NESSUS to look for vulnerabilities in your network. These tools execute port scans as part of their testing. Amazon and some other cloud providers specifically prohibit this kind of activity, and you can end up violating your terms of service.

Implementing network intrusion detection in the cloud

As I mentioned in the previous section, you simply cannot implement a network intrusion detection system in the Amazon cloud (or any other cloud that does not expose LAN traffic) that passively listens to local network traffic. Instead, you must run the NIDS on your load balancer or on each server in your infrastructure. There are advantages and disadvantages to each approach, but I am not generally a fan of NIDS in the cloud unless required by a standard or regulation.

The simplest approach is to have a dedicated NIDS server in front of the network as a whole that watches all incoming traffic and acts accordingly. Figure 5-4 illustrates this architecture.

Because the only software running on the load balancer is the NIDS software and Apache, it maintains a very low attack profile. Compromising the NIDS server requires a vulnerability in

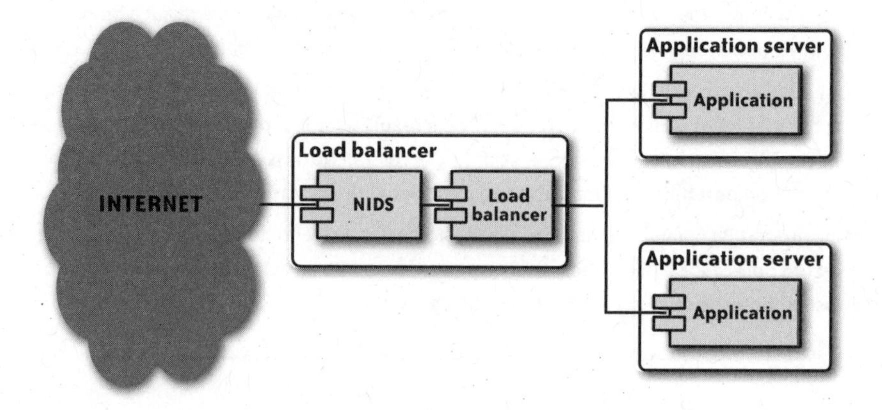

FIGURE 5-4. A network intrusion detection system listening on a load balancer

the NIDS software or Apache—assuming the rest of the system is properly hardened and no actual services are listening to any other ports open to the Web as a whole.

The load balancer approach creates a single point of failure for your network intrusion detection system because, in general, the load balancer is the most exposed component in your infrastructure. By finding a way to compromise your load balancer, the intruder not only takes control of the load balancer, but also has the ability to silence detection of further attacks against your cloud environment.

You can alternately implement intrusion detection on a server behind the load balancer that acts as an intermediate point between the load balancer and the rest of the system. This design is generally superior to the previously described design, except that it leaves the load balancer exposed (only traffic passed by the load balancer is examined) and reduces the overall availability of the system.

Another approach is to implement network intrusion detection on each server in the network. This approach creates a very slight increase in the attack profile of the system as a whole because you end up with common software on all servers. A vulnerability in your NIDS would result in a vulnerability on each server in your cloud architecture. On a positive note, you make it much more difficult for an intruder to hide his footprints.

As I mentioned earlier, I am not a huge fan of network intrusion detection in the Amazon cloud.† Unlike a traditional infrastructure, there just is no meaningful way for a NIDS to serve its purpose. You simply cannot devise any NIDS architecture that will give your NIDS visibility to all traffic attempting to reach your instances. The best you can do is create an implementation in which the NIDS is deployed on each server in your infrastructure with visibility to the traffic that Amazon allows into the security group in which the instance is deployed. You would see

† Let me strongly repeat that qualification: *in the Amazon cloud*. Your data center and most other clouds will benefit from a network intrusion detection system.

minimally valid proactive alerting, and the main benefit would be protection against malicious payloads. But, if you are encrypting all your traffic, even that benefit is minimal. On the other hand, the presence of a NIDS will greatly reduce the performance of those servers and create a single attack vector for all hosts in your infrastructure.

Host Security

Host security describes how your server is set up for the following tasks:

- Preventing attacks.
- Minimizing the impact of a successful attack on the overall system.
- Responding to attacks when they occur.

It always helps to have software with no security holes. Good luck with that! In the real world, the best approach for preventing attacks is to assume your software has security holes. As I noted earlier in this chapter, each service you run on a host presents a distinct attack vector into the host. The more attack vectors, the more likely an attacker will find one with a security exploit. You must therefore minimize the different kinds of software running on a server.

Given the assumption that your services are vulnerable, your most significant tool in preventing attackers from exploiting a vulnerability once it becomes known is the rapid rollout of security patches. Here's where the dynamic nature of the cloud really alters what you can do from a security perspective. In a traditional data center, rolling out security patches across an entire infrastructure is time-consuming and risky. In the cloud, rolling out a patch across the infrastructure takes three simple steps:

1. Patch your AMI with the new security fixes.
2. Test the results.
3. Relaunch your virtual servers.

Here a tool such as enStratus or RightScale for managing your infrastructure becomes absolutely critical. If you have to manually perform these three steps, the cloud can become a horrible maintenance headache. Management tools, however, can automatically roll out the security fixes and minimize human involvement, downtime, and the potential for human-error-induced downtime.

System Hardening

Prevention begins when you set up your machine image. As you get going, you will experiment with different configurations and constantly rebuild images. Once you have found a configuration that works for a particular service profile, you should harden the system before creating your image.

Server hardening is the process of disabling or removing unnecessary services and eliminating unimportant user accounts. Tools such as Bastille Linux can make the process of hardening your machine images much more efficient. Once you install Bastille Linux, you execute the interactive scripts that ask you questions about your server. It then proceeds to disable services and accounts. In particular, it makes sure that your hardened system meets the following criteria:

- No network services are running except those necessary to support the server's function.
- No user accounts are enabled on the server except those necessary to support the services running on the server or to provide access for users who need it.
- All configuration files for common server software are configured to the most secure settings.
- All necessary services run under a nonprivileged role user account (e.g., run MySQL as the mysql user, not root).
- When possible, run services in a restricted filesystem, such as a chroot jail.

Before bundling your machine image, you should remove all interactive user accounts and passwords stored in configuration files. Although the machine image will be stored in an encrypted format, Amazon holds the encryption keys and thus can be compelled to provide a third party with access through a court subpoena.

Antivirus Protection

Some regulations and standards require the implementation of an antivirus (AV) system on your servers. It's definitely a controversial issue, since an AV system with an exploit is itself an attack vector and, on some operating systems, the percentage of AV exploits to known viruses is relatively high.

Personally, I have mixed feelings about AV systems. They are definitely necessary in some circumstances, but a risk in others. For example, if you are accepting the upload of photos or other files that could be used to deliver viruses that are then served to the public, you have an obligation to use some kind of antivirus software in order to protect your site from becoming a mechanism for spreading the virus.

Unfortunately, not all AV systems are created equally. Some are written better than others, and some protect you much better than others. Finally, some servers simply don't have an operational profile that makes viruses, worms, and trojans viable attack vectors. I am therefore bothered by standards, regulations, and requirements that demand blanket AV coverage.

When looking at the AV question, you first should understand what your requirements are. If you are required to implement AV, then you should definitely do it. Look for two critical features in your AV software:

- How wide is the protection it provides? In other words, what percentage of known exploits does it cover?‡
- What is the median delta between the time when a virus is released into the wild and the time your AV product of choice provides protection against it?

Once you have selected an AV vendor and implemented it on your servers, you absolutely must keep your signatures up to date. You are probably better off with no AV system than one with outdated versions or protections.

Host Intrusion Detection

Whereas a network intrusion detection system monitors network traffic for suspicious activity, a *host intrusion detection system* (HIDS) such as OSSEC monitors the state of your server for anything unusual. An HIDS is in some ways similar to an AV system, except it examines the system for all signs of compromise and notifies you when any core operating system or service file changes.

> **WARNING**
>
> **I may be ambivalent about NIDS and AV in the cloud, but I am a strong proponent of HIDS in the cloud. You should not deploy servers in the cloud without an HIDS.**

In my Linux deployments, I use OSSEC (*http://www.ossec.net*) for host-based intrusion detection. OSSEC has two configuration profiles:

- Standalone, in which each server scans itself and sends you alerts.
- Centralized, in which you create a centralized HIDS server to which each of the other servers sends reports.

In the cloud, you should always opt for the centralized configuration. It centralizes your rules and analysis so that it is much easier to keep your HIDS infrastructure up to date. Furthermore, it enables you to craft a higher security profile for your HIDS processing than the individual services might allow for. Figure 5-5 illustrates a cloud network using centralized HIDS.

As with an AV solution, you must keep your HIDS servers up to date constantly, but you do not need to update your individual servers as often.

The downside of an HIDS is that it requires CPU power to operate, and thus can eat up resources on your server. By going with a centralized deployment model, however, you can push a lot of that processing onto a specialized intrusion detection server.

‡ So many viruses are created these days that most AV companies cannot possibly develop protection against all of them. They knowingly punt on protecting you against published viruses.

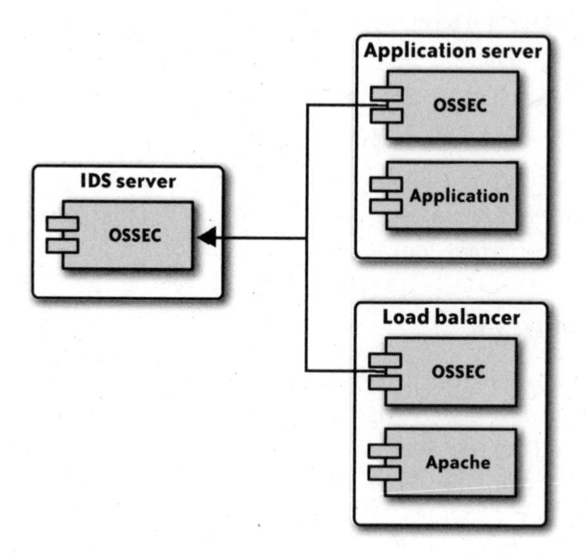

FIGURE 5-5. A HIDS infrastructure reporting to a centralized server

Data Segmentation

In addition to assuming that the services on your servers have security exploits, you should further assume that eventually one of them will be compromised. Obviously, you never want any server to be compromised. The best infrastructure, however, is tolerant of—in fact, it assumes—the compromise of any individual node. This tolerance is not meant to encourage lax security for individual servers, but is meant to minimize the impact of the compromise of specific nodes. Making this assumption provides you with a system that has the following advantages:

- Access to your most sensitive data requires a full system breach.
- The compromise of the entire system requires multiple attack vectors with potentially different skill sets.
- The downtime associated with the compromise of an individual node is negligible or nonexistent.

The segmentation of data based on differing levels of sensitivity is your first tool in minimizing the impact of a successful attack. We examined a form of data segmentation in Chapter 4 when we separated credit card data from customer data. In that example, an attacker who accesses your customer database has found some important information, but that attacker still lacks access to the credit card data. To be able to access credit card data, decrypt it, and associate it with a specific individual, the attacker must compromise both the e-commerce application server and the credit card processor.

Here again the approach of one server/one service helps out. Because each type of server in the chain offers a different attack vector, an attacker will need to exploit multiple attack vectors to compromise the system as a whole.

Credential Management

Your machine images OSSEC profileshould have no user accounts embedded in them. In fact, you should never allow password-based shell access to your virtual servers. The most secure approach to providing access to virtual servers is the dynamic delivery of public SSH keys to target servers. In other words, if someone needs access to a server, you should provide her credentials to the server when it starts up or via an administrative interface instead of embedding that information in the machine image.

> **NOTE**
>
> **The best credential management system provides remote access to the server only for users who absolutely have an operational need to access the server and only for the time period for which they need that access. I strongly recommend taking advantage of the dynamic nature of the cloud to eliminate the need to keep access credentials for people with little or no current need to access your server.**

Of course, it is perfectly secure to embed public SSH keys in a machine image, and it makes life a lot easier. Unfortunately, it makes it harder to build the general-purpose machine images I described in Chapter 4. Specifically, if you embed the public key credentials in a machine image, the user behind those credentials will have access to every machine built on that image. To remove her access or add access for another individual, you subsequently have to build a new machine image reflecting the changed dynamics.

Therefore, you should keep things simple and maintainable by passing in user credentials as part of the process of launching your virtual server. At boot time, the virtual server has access to all of the parameters you pass in and can thus set up user accounts for each user you specify. It's simple because it requires no tools other than those that Amazon already provides. On the other hand, adding and removing access after the system boots up becomes a manual task.

Another approach is to use existing cloud infrastructure management tools or build your own that enable you to store user credentials outside the cloud and dynamically add and remove users to your cloud servers at runtime. This approach, however, requires an administrative service running on each host and thus represents an extra attack vector against your server.

Compromise Response

Because you should be running an intrusion detection system, you should know very quickly if and when an actual compromise occurs. If you respond rapidly, you can take advantage of the cloud to eliminate exploit-based downtime in your infrastructure.

When you detect a compromise on a physical server, the standard operating procedure is a painful, manual process:

1. Remove intruder access to the system, typically by cutting the server off from the rest of the network.
2. Identify the attack vector. You don't want to simply shut down and start over, because the vulnerability in question could be on any number of servers. Furthermore, the intruder very likely left a rootkit or other software to permit a renewed intrusion after you remove the original problem that let him in. It is therefore critical to identify how the intruder compromised the system, if that compromise gave him the ability to compromise other systems, and if other systems have the same vulnerability.
3. Wipe the server clean and start over. This step includes patching the original vulnerability and rebuilding the system from the most recent *uncompromised* backup.
4. Launch the server back into service and repeat the process for any server that has the same attack vector.

This process is very labor intensive and can take a long time. In the cloud, the response is much simpler.

First of all, the forensic element can happen after you are operating. You simply copy the root filesystem over to one of your block volumes, snapshot your block volumes, shut the server down, and bring up a replacement.

Once the replacement is up (still certainly suffering from the underlying vulnerability, but at least currently uncompromised), you can bring up a server in a dedicated security group that mounts the compromised volumes. Because this server has a different root filesystem and no services running on it, it is not compromised. You nevertheless have full access to the underlying compromised data, so you can identify the attack vector.

With the attack vector identified, you can apply patches to the machine images. Once the machine images are patched, simply relaunch all your instances. The end result is a quicker response to a vulnerability with little (if any) downtime.

CHAPTER SIX

Disaster Recovery

HOW GOOD IS YOUR DISASTER RECOVERY PLAN? It's fully documented and you regularly test it by running disaster recovery drills, right?

So far in this book, we have talked about what happens in the event of routine, expected failures. Disaster recovery is the practice of making a system capable of surviving unexpected or extraordinary failures. A disaster recovery plan, for example, will help your IT systems survive a fire in your data center that destroys all of the servers in that data center and the systems they support.

Every organization should have a documented disaster recovery process and should test that process at least twice each year. In reality, even well-disciplined companies tend to fall short in their disaster recovery planning. Too many small- and medium-size businesses would simply go out of business in the case of the data center fire scenario I just outlined.

One of the things I personally love about virtualization is the way it lets you automate disaster recovery. Recovery from trivial failures and disaster recovery in the cloud become largely indistinguishable operations. As a result, if your entire cloud infrastructure falls apart, you should have the capabilities in place to restore it on internal servers, at a managed hosting services provider, or at another cloud provider in minutes or hours.

Disaster Recovery Planning

Disaster recovery deals with catastrophic failures that are extremely unlikely to occur during the lifetime of a system. If they are reasonably expected failures, they fall under the auspices of traditional availability planning. Although each single disaster is unexpected over the lifetime of a system, the possibility of some disaster occurring over time is reasonably nonzero.

Through disaster recovery planning, you identify an acceptable recovery state and develop processes and procedures to achieve the recovery state in the event of a disaster. By "acceptable recovery state," I specifically mean how much data you are willing to lose in the event of a disaster.

Defining a disaster recovery plan involves two key metrics:

Recovery Point Objective (RPO)
: The recovery point objective identifies how much data you are willing to lose in the event of a disaster. This value is typically specified in a number of hours or days of data. For example, if you determine that it is OK to lose 24 hours of data, you must make sure that the backups you'll use for your disaster recovery plan are never more than 24 hours old.

Recovery Time Objective (RTO)
: The recovery time objective identifies how much downtime is acceptable in the event of a disaster. If your RTO is 24 hours, you are saying that up to 24 hours may elapse between the point when your system first goes offline and the point at which you are fully operational again.

In addition, the team putting together a disaster recovery plan should define the criteria that would trigger invocation of the plan. In general, invocation of any plan that results in accepting a loss of data should involve the heads of the business organization—even if the execution of the plan is automated, as I am promoting in this chapter.

Everyone would love a disaster recovery scenario in which no downtime and no loss of data occur, no matter what the disaster. The nature of a disaster, however, generally requires you to accept some level of loss; anything else will come with a significant price tag. In a citywide disaster like Hurricane Katrina, the cost of surviving with zero downtime and zero data loss could have been having multiple data centers in different geographic locations that were constantly synchronized. In other words, you would need two distinct data centers from different infrastructure providers with dedicated, high-bandwidth connections between the two.

Accomplishing that level of redundancy is expensive. It would also come with a nontrivial performance penalty. The cold reality for most businesses is likely that the cost of losing 24 hours of data is less than the cost of maintaining a zero downtime/zero loss of data infrastructure.

Determining an appropriate RPO and RTO is ultimately a financial calculation: at what point does the cost of data loss and downtime exceed the cost of a backup strategy that will prevent that level of data loss and downtime? The right answer is radically different for different businesses. If you are in a senior IT management role in an organization, you should definitely know the right answer for your business.

The final element of disaster recovery planning is understanding the catastrophic scenario. There's ultimately some level of disaster your IT systems will not survive no matter how much planning and spending you do. A good disaster recovery plan can describe that scenario so that all stakeholders can understand and accept the risk.

The Recovery Point Objective

The easiest place to start is your RPO. The Armageddon scenario results in total loss of all system data and the binaries of all applications required to run the system. Your RPO is somewhere between the application state when you first deployed it and the state at the time of the disaster. You may even define multiple disaster levels with different RPOs.*

Just about any software system should be able to attain an RPO between 24 hours for a simple disaster to one week for a significant disaster without incurring absurd costs. Of course, losing 24 hours of banking transactions would never be acceptable, much less one week.

Your RPO is typically governed by the way in which you save and back up data:

- Weekly off-site backups will survive the loss of your data center with a week of data loss. Daily off-site backups are even better.
- Daily on-site backups will survive the loss of your production environment with a day of data loss plus replicating transactions during the recovery period after the loss of the system. Hourly on-site backups are even better.
- A NAS/SAN will survive the loss of any individual server, except for instances of data corruption with no data loss.
- A clustered database will survive the loss of any individual data storage device or database node with no data loss.
- A clustered database across multiple data centers will survive the loss of any individual data center with no data loss.

Later in this chapter, we talk about how the cloud changes your options and makes lower RPOs possible.

The Recovery Time Objective

Having up-to-the-second off-site backups does you no good if you have no environment to which you can restore them in the event of failure. The ability to assemble a replacement infrastructure for your disasters—including the data restore time—governs the RTO.

* You might, for example, define a level 1 disaster as involving the loss of a single data center and a level 2 disaster as involving the loss of multiple data centers.

What would happen if your managed services provider closed its doors tomorrow? If you have a number of dedicated servers, it can be days or weeks before you are operational again unless you have an agreement in place for a replacement infrastructure.†

In a traditional infrastructure, a rapid RTO is very expensive. As I already noted, you would have to have an agreement in place with another managed services provider to provide either a backup infrastructure or an SLA for setting up a replacement infrastructure in the event your provider goes out of business. Depending on the nature of that agreement, it can nearly double the costs of your IT infrastructure.

The cloud—even over virtualized data centers—alters the way you look at your RTO. We'll dive into the reasons why later in the chapter.

Disasters in the Cloud

Assuming unlimited budget and capabilities, I focus on three key things in disaster recovery planning:

1. Backups and data retention
2. Geographic redundancy
3. Organizational redundancy

If I can take care of those three items, it's nearly certain I can meet most RPO and RTO needs. But I have never been in a situation in which I had an unlimited budget and capabilities, so I have always had to compromise. As a result, the order of the three items matters. In addition, if your hosting provider is a less-proven organization, organizational redundancy may be more important than geographic redundancy.

Fortunately, the structure of the Amazon cloud makes it very easy to take care of the first and second items. In addition, cloud computing in general makes the third item much easier.

> **NOTE**
>
> **The section contains a lot of content that is specific to the Amazon cloud. Most clouds currently have or are developing concepts similar to EC2 block storage devices and snapshots. For such clouds, many of these concepts apply. Issues relating to geographic redundancy, however, are specific to your cloud provider, so very little that I say on that subject applies to other cloud providers.**

† This scenario also illustrates that simply having off-site backups is not enough. Those backups must reside outside the control of your managed services provider, or you risk losing access to them in the event of that provider going bankrupt.

Backup Management

In Chapter 4, I looked at the technical details of how to manage AMIs and execute backups in a cloud environment. Now it's time to take a step back from the technical details and examine the kinds of data you are planning to back up and how it all fits into your overall disaster recovery plan.

Your ability to recover from a disaster is limited by the quality and frequency of your backups. In a traditional IT infrastructure, companies often make full weekly backups to tape with nightly differentials and then ship the weekly backups off-site. You can do much better in the cloud, and do it much more cheaply, through a layered backup strategy.

BACKUPS, BUSINESS CONTINUITY, AND AWS

In this section, I cover a number of technologies that Amazon Web Services provide to help you manage backups effectively. If you are using a different cloud, they likely have some similar tools as well as some that are completely different. A critical part of any backup strategy, however, is the concept of off-site backups. Whatever your backup strategy, you must not only have a mechanism for moving all data critical for achieving your Recovery Point Objectives out of the cloud, but you must also store that data in a portable format so you can recover into an environment that might be radically different from your current cloud provider.

Table 6-1 illustrates the different kinds of data that web applications typically manage.

TABLE 6-1. Backup requirements by data type

Kind of data	Description
Fixed data	Fixed data, such as your operating system and common utilities, belong in your AMI. In the cloud, you don't back up your AMI, because it has no value beyond the cloud.[a]
Transient data	File caches and other data that can be lost completely without impacting the integrity of the system. Because your application state is not dependent on this data, don't back it up.
Configuration data	Runtimeconfiguration data necessary to make the system operate properly in a specific context. This data is not transient, since it must survive machine restarts. On the other hand, it should be easily reconfigured from a clean application install. This data should be backed up semi-regularly.
Persistent data	Your application state, including critical customer data such as purchase orders. It changes constantly and a database engine is the best tool for managing it. Your database engine should store its state to a block device, and you should be performing constant backups. Clustering and/or replication are also critical tools in managing the database.

[a] Keep in mind that even if Amazon S3 failed completely and lost your AMIs, as long as EC2 is available and you have EC2 instances running based on the lost AMI, you will be able to quickly rebuild the lost AMI. On the other hand, if EC2 goes down, S3 goes down, and all S3 data is lost completely, you'll have to recover into a different cloud that doesn't recognize your AMI anyway!

In disaster recovery, persistent data is generally the data of greatest concern. We can always rebuild the operating system, install all the software, and reconfigure it, but we have no way of manually rebuilding the persistent data.

Fixed data strategy

If you are fixated on the idea of backing up your machine images, you can download the images out of S3 and store them outside of the Amazon cloud. If S3 were to go down and incur data loss or corruption that had an impact on your AMIs, you would be able to upload the images from your off-site backups and reregister them. It's not a bad idea and it is not a lot of trouble, but the utility is limited given the uniqueness of the failure scenario that would make you turn to those backups.

Configuration data strategy

A good backup strategy for configuration information comprises two levels. The first level can be either a regular filesystem dump to your cloud storage or a filesystem snapshot. For most applications, you can back up your configuration data once a day or even once a week and be fine. You should, however, think back to your Recovery Point Objective. If your configuration data changes twice a day and you have a two-hour RPO, you will need to back up your configuration data twice a day. If configuration data changes irregularly, it may be necessary to make hourly backups or specifically tie your backups to changes in application configuration.

An alternate approach is to check your application configuration into a source code repository outside of the cloud and leverage that repository for recovery from even minor losses.

Whether you perform filesystem snapshots or simply zip up the filesystem, that data will hibernate inside S3. Snapshots tend to be the most efficient and least intrusive mechanism for performing backups, but they are also the least portable. You don't have direct access to the EC2 snapshots, and even if you did, they would not be usable outside of the Amazon cloud. At some point, you do need to get that data out of the cloud so that you have off-site backups in a portable format. Here's what I recommend:

- Create regular—at a minimum, daily—snapshots of your configuration data.
- Create semi-regular—at least less than your RPO—filesystem archives in the form of ZIP or TAR files and move those archives into Amazon S3.
- On a semi-regular basis—again, at least less than your RPO—copy your filesystem archives out of the Amazon cloud into another cloud or physical hosting facility.

Let's say I had a one-day RPO with an application housing less than 10 GB of data whose configuration data could change on a whim:

- I would make hourly snapshots of the filesystem with the configuration data and archive the whole thing in a portable format to S3 at least once a day.

- I would retain a week's worth of full backups in S3 in case of a corruption issue during backup. In addition, I would copy each daily backup out of S3 to another cloud or my internal file servers at some point right after it is made.
- Finally, I would keep a week's worth of backups off site for the last week along with one backup a week for the prior month and one backup a month for the prior year.

Depending on the amount of data in question and the data retention issues/requirements, I might do more or less in the archiving of old backups.

Different RPOs and application behaviors mandate different strategies. I have to emphasize again: you must absolutely understand your RPO in order to create a proper backup strategy. Given the example strategy I just outlined, I will be able to create a functional replication of my application in any data center in the world and lose at most 24 hours of data, except in the event of complete data corruption of backups or simultaneous loss of all of Amazon and wherever my off-site backups are stored.

Persistent data strategy (aka database backups)

I've already recommended using a relational database to store customer information and other persistent data. After all, the purpose of a relational database is to maintain the consistency of complex transactional data. The challenge of setting up database backups is doing them regularly in a manner that does not impact operations while retaining database integrity.

MySQL, like all database engines, provides several convenient tools for backups, but you must use them carefully to avoid data corruption. The techniques are fairly well-documented in the database literature, but they're important enough for me to summarize them here as well as apply them to an Amazon EC2 environment.

If you aren't familiar with the way databases work, you might be wondering, "Why is that so hard? Why don't I do a snapshot or make an archive like I did with the configuration data?" With the configuration data, it is highly unlikely you will be making a backup in between the writing of two different files that must remain consistent or in the middle of writing out a file to the filesystem. With database storage, it is a near certainty that every time you try to copy those files, the database will be in the middle of doing something with them. As a result, you need to get clever with your database backup strategy.

The first line of defense is either *multimaster replication* or *clustering*. A multimaster database is one in which two master servers execute write transactions independently and replicate the transactions to the other master. A clustered database environment contains multiple servers that act as a single logical server. Under both scenarios, when one goes down, the system remains operational and consistent.

Instead, you can perform *master-slave replication*. Master-slave replication involves setting up a *master server* that handles your write operations and replicating transactions over to a *slave server*. Each time something happens on the master, it replicates to the slave.

Replication in itself is not a "first line of defense," since replication is not atomic with respect to the transactions that take place on the master. In other words, a master can crash after a transaction has completed but before it has had time to replicate to the slave. To get around this problem, I generally do the following:

- Set up a master with its data files stored on a block storage device.
- Set up a replication slave, storing its data files on a block storage device.
- Take regular snapshots of the master block storage device based on my RPO.
- Create regular database dumps of the slave database and store them in S3.
- Copy the database dumps on a semi-regular basis from S3 to a location outside the Amazon cloud.

> **NOTE**
>
> **When dumping your MySQL database state to offline media, I strongly encourage doing so off a replication slave that is not supporting read-only queries for the live application. By leveraging the slave, you minimize the impact of performing backups on the production system.**

Amazon's Elastic Block Storage (EBS) offering has not been around long enough to provide a good feel for how likely you are to see database corruption in the event of your MySQL server failing. I have taken the position, however, that corruption is extremely likely—as it is with many filesystems—and have adjusted my backup strategy accordingly. I could be proven wrong on this issue, but for the moment, I prefer to play it safe.

Actually taking snapshots or creating database dumps for some database engines is actually very tricky in a runtime environment, especially if you want to do it hourly or even more frequently. The challenge in creating your backups for these database engines is the need to stop processing all transactions while the backup is taking place. To complicate the situation, database dumps can take a long time to complete. As a result, your applications will grind to a halt while you make any database dumps.

Snapshots are available in most cloud environments and provide an important approach for maintaining database integrity without completely shutting down application processing—even with large data sets in databases such as MySQL.

You need to freeze the database only for an instant to create your snapshot. The process follows these steps:

1. Lock the database.
2. Sync the filesystem (this procedure is filesystem-dependent).
3. Take a snapshot.
4. Unlock the database.

The whole process should take about one second.

On Amazon EC2, you will store your snapshots directly onto block storage. Unfortunately, the snapshots are not portable, so you can't use them for off-site storage. You therefore will need to do database dumps, no matter how much you would rather avoid doing them. Because of this need, I run my backups against a database slave. The slave can afford to be locked for a period of time while a database dump completes. The fact that the slave may be a transaction or two (or ten) behind the master is unimportant for the purposes of making a backup. What matters is that you can create the dump without impacting your applications.

NOTE

With very large databases, performing dumps may simply take too long to be feasible. An alternate approach involves backing up the database dumps along with the database logfiles so that they, too, can be used to restore data after a crash or mistaken change. Alternately, with MySQL, you can simply back up the database files. Both approaches are slightly more complex than a straight database dump.

The steps for creating the database dump are:

1. Execute the database dump.
2. When complete, encrypt the dump and break it into small, manageable chunks.
3. Move the dump over to S3.

Amazon S3 limits your file size to 5 GB. As a result, you probably need to break your database into chunks, and you should definitely encrypt it and anything else you send into Amazon S3.

Now that you have a portable backup of your database server, you can copy that backup out of the Amazon cloud and be protected from the loss of your S3 backups.

Backup security

If you are following the security practices I outlined earlier in the book, your filesystems are encrypted to protect the snapshots you are making for your backups from prying eyes. The harder part is securing your portable backups as you store them in S3 and move them off site.

I typically use PGP-compatible encryption for my portable backups. You need to worry about two issues:

- Keeping your private decryption key out of the cloud.
- Keeping your private decryption key some place that it will never, ever get lost.

Given the disaster recovery procedures I cover in this chapter, you really have no reason for ever giving out the private decryption key to an instance in the cloud unless you are automating the failover between two different cloud infrastructures. The cloud needs only your public encryption key so it can encrypt the portable backups.

You can't store your decryption key with your backups. Doing so will defeat the purpose of encrypting the backups in the first place. Because you will store your decryption key somewhere else, you run the risk of losing your decryption key independent of your backups. On the other hand, keeping a bunch of copies of your decryption key will make it more likely it will fall into the wrong hands.

The best approach? Keep two copies:

- One copy stored securely on a highly secure server in your internal network.
- One copy printed out on a piece of paper and stored in a safety deposit box nowhere near the same building in which you house your highly secure server.

More than one person should know the locations of these copies. A true disaster can unfortunately result in the loss of personnel, so personnel redundancy is also important for a disaster recovery plan.

If you are automating the recovery from portable backups, you will also need to keep a copy of the private decryption key on the server that orchestrates your automated recovery efforts.

Geographic Redundancy

Everything I have discussed so far in this chapter focuses on the Recovery Point Objective. In the end, that's the easiest part of disaster recovery. The virtualization technologies behind the cloud simply make it a lot easier to automate those processes and have a relatively inexpensive mechanism for off-site backups.

Turning now to your Recovery Time Objective, the key is redundancy in infrastructure. If you can develop geographical redundancy, you can survive just about any physical disaster that might happen. With a physical infrastructure, geographical redundancy is expensive. In the cloud, however, it is relatively cheap.

You don't necessarily need to have your application running actively in all locations, but you need the ability to bring your application up from the redundant location in a state that meets your Recovery Point Objective within a timeframe that meets your Recovery Time Objective. If you have a 2-hour RTO with a 24-hour RPO, geographical redundancy means that your second location can be operational within two hours of the complete loss of your primary location using data that is no older than 24 hours.

Amazon provides built-in geographic redundancy in the form of regions and availability zones. If you have your instances running in a given availability zone, you can get them started back up in another availability zone in the same region without any effort. If you have specific requirements around what constitutes geographic redundancy,‡ Amazon's availability zones may not be enough—you may have to span regions.

‡ Some government agencies and other organizations mandate a separation of at least 50 miles.

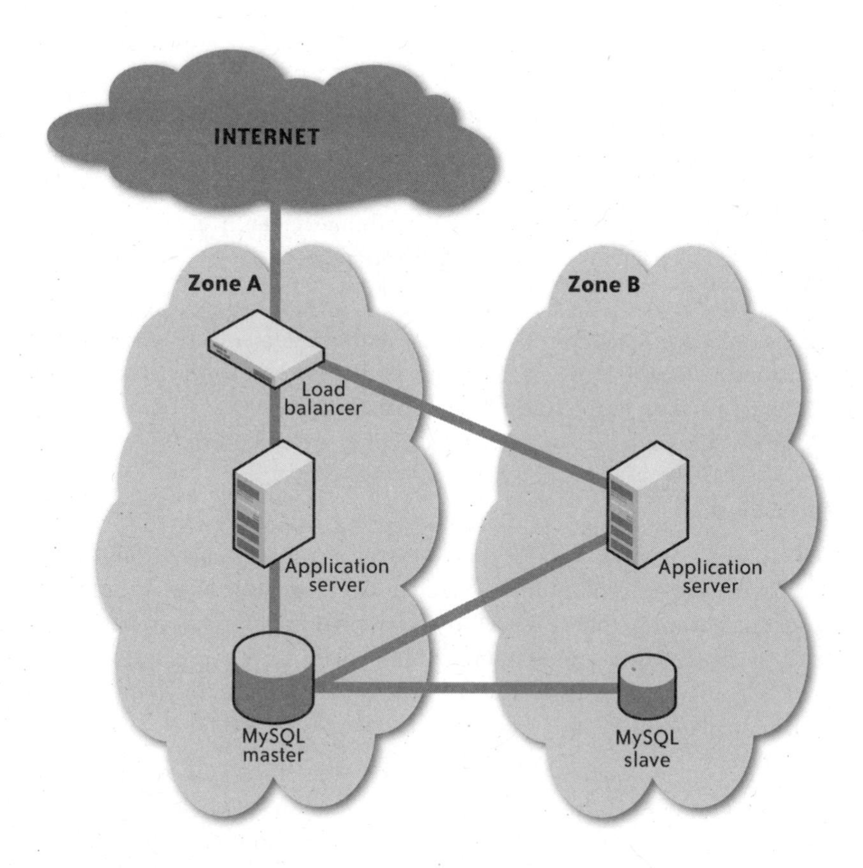

FIGURE 6-1. By spanning multiple availability zones, you can achieve geographic redundancy

Spanning availability zones

Just about everything in your Amazon infrastructure except block storage devices is available across all availability zones in a given region. Although there is a charge for network traffic that crosses availability zones, that charge is generally well worth the price for the leveraging ability to create redundancy across availability zones.

Figure 6-1 illustrates an application environment that can easily tolerate the loss of an entire availability zone.

If you lose the entire availability zone B, nothing happens. The application continues to operate normally, although perhaps with degraded performance levels.

If you lose availability zone A, you will need to bring up a new load balancer in availability zone B and promote the slave in that availability zone to master. The system can return to operation in a few minutes with little or no data loss. If the database server were clustered and you had a spare load balancer running silently in the background, you could reassign the IP address from the old load balancer to the spare and see only a few seconds of downtime with no data loss.

The Amazon SLA provides for a 99.95% uptime of at least two availability zones in each region. If you span multiple availability zones, you can actually exceed the Amazon SLA in regions that have more than two availability zones. The U.S. East Coast, for example, has three availability zones.[§] As a result, you have only a 33% chance of any given failure of two availability zones being exactly the two zones you are using.

Even in the event that you are unfortunate enough to be operating in exactly the two zones that fail, you can still exceed Amazon's SLA as long as the region you are operating in has more than two availability zones. The trick is to execute your disaster recovery procedures and bring your infrastructure back up in the remaining availability zone. As a result, you can be operational again while the other two availability zones are still down.

Operating across regions

At the time I write this chapter, Amazon supports two regions: us-east-1 (Eastern United States) and eu-west-1 (Western Europe). These regions share little or no meaningful infrastructure. The advantage of this structure is that your application can basically survive a nuclear attack on the U.S. or EU (but not on both!) if you operate across regions. On the other hand, the lack of common infrastructure makes the task of replicating your environments across regions more difficult.

Each region has its own associated Amazon S3 region. Therefore, you cannot launch EC2 instances in the EU using AMIs from the U.S., and you cannot use IP addresses formerly associated with a load balancer in the EU with a replacement in the U.S.

> **NOTE**
>
> **As I write this, the whole idea of multiple regions is new to AWS. It is possible that some of the restrictions I mention here are no longer true as you read this book. Furthermore, as people gain more experience operating across multiple regions, they are likely to develop best practices I am not considering in this section.**

How you manage operations across regions depends on the nature of your web application and your redundancy needs. It's entirely likely that just having the capability to rapidly launch in another region is good enough, without actually developing an infrastructure that simultaneously operates in both regions.

The issues you need to consider for simultaneous operation include:

§ The EU region has exactly two availability zones. As a result, the best uptime you can promise in the EU region alone is 99.95% availability.

DNS management

You can use round-robin DNS to work around the fact that IP addresses are not portable across regions, but you will end up sending European visitors to the U.S. and vice versa (very inefficient network traffic management) and lose half your traffic when one of the regions goes down. You can leverage a dynamic DNS system such as UltraDNS that will offer up the right DNS resolution based on source and availability.

Database management

Clustering across regions is likely not practical (but you can try it). You can also set up a master in one region with a slave in the other. Then you perform write operations against the master, but read against the slave for traffic from the region with the slave. Another option is to segment your database so that the European region has "European data" and the U.S. region has "American data." Each region also has a slave in the other region to act as a recovery point from the full loss of a region.

Regulatory issues

The EU does not allow the storage of certain data outside of the EU. As a result, legally you may not be allowed to operate across regions, no matter what clever technical solutions you devise. In reality, an Amazon+GoGrid or Amazon+Rackspace approach to redundancy may be more effective than trying to use Amazon's two cross-jurisdictional regions.

For most purposes, I recommend a process for regularly copying infrastructure elements (AMIs, backups, and configuration) over into the other region and then having the ability to rapidly start that infrastructure in the event of a total, prolonged failure of your core zone.

Organizational Redundancy

If you have an infrastructure that does everything I have recommended so far, you are pretty well protected against everything physical that can happen. You are still exposed at the business level, however. Specifically, if Amazon or Rackspace or GoGrid or whoever you are using goes out of business or decides it is bored with cloud computing, you might find yourself in trouble.

Physical disasters are a relatively rare thing, but companies go out of business everywhere every day—even big companies like Amazon and Rackspace. Even if a company goes into bankruptcy restructuring, there's no telling what will happen to the hardware assets that run their cloud infrastructure. Your disaster recovery plan should therefore have contingencies that assume your cloud provider simply disappears from the face of the earth.

You probably won't run concurrent environments across multiple clouds unless it provides some level of geographic advantage. Even in that case, your environments are not likely to be redundant so much as segmented for the geographies they are serving. Instead, the best approach to organizational redundancy is to identify another cloud provider and establish a backup environment with that provider in the event your first provider fails.

WARNING

When selecting an alternative cloud provider as your backup provider, you must verify that this provider does not somehow rely on your primary provider for anything. For example, if your secondary cloud provider uses the data center of your primary cloud provider to host their physical infrastructure, you won't be protected against the failure of your primary provider.

The issues associated with organizational redundancy are similar to the issues I discussed earlier around operating across Amazon EC2 regions. In particular, you must consider all of the following concerns:

- Storing your portable backups at your secondary cloud provider.
- Creating machine images that can operate your applications in the secondary provider's virtualized environment.
- Keeping the machine images up to date with respect to their counterparts with the primary provider.
- Not all cloud providers and managed service providers support the same operation systems or filesystems. If your application is dependent on either, you need to make sure you select a cloud provider that can support your needs.

Disaster Management

You are performing your backups and have an infrastructure in place with all of the appropriate redundancies. To complete the disaster recovery scenario, you need to recognize when a disaster has happened and have the tools and processes in place to execute your recovery plan. One of the coolest things about the cloud is that all of this can be automated. You can recover from the loss of Amazon's U.S. data centers while you sleep.[‖]

Monitoring

Monitoring your cloud infrastructure is extremely important. You cannot replace a failing server or execute your disaster recovery plan if you don't know that there has been a failure. The trick, however, is that your monitoring systems cannot live in either your primary or secondary cloud provider's infrastructure. They must be independent of your clouds. If you want to enable automated disaster recovery, they also need the ability to manage your EC2 infrastructure from the monitoring site.

‖ Though it is probably not a good idea to automatically engage disaster recovery processes without some human intervention when data could lost during the launch of those processes.

NOTE

There exist a number of tools, such as enStratus and RightScale, that manage your infrastructure for you. Some even automate your disaster recovery processes so you don't have to write any custom code.

Your primary monitoring objective should be to figure out what is going to fail before it actually fails. The most common problem I have encountered in EC2 is servers that gradually decrease in local file I/O throughput until they become unusable. This problem is something you can easily watch for and fix before users even notice it. On the other hand, if you wait for your application to fail, chances are users have had to put up with poor performance for some period of time before it failed completely. It may also prove to be a precursor to a larger cloud failure event.

There are many other more mundane things that you should check on in a regular environment. In particular, you should be checking capacity issues such as disk usage, RAM, and CPU. I cover those things in more detail in Chapter 7.

In the end, however, you will need to monitor for failure at three levels:

- Through the provisioning API (for Amazon, the EC2 web services API)
- Through your own instance state monitoring tools
- Through your application health monitoring tools

Your cloud provider's provisioning API will tell you about the health of your instances, any volumes they are mounting, and the data centers in which they are operating. When you detect a failure at this level, it likely means something has gone wrong with the cloud itself. Before engaging in any disaster recovery, you will need to determine whether the outage is limited to one server or affects indeterminate servers, impacting an entire availability zone or an entire region.

Monitoring is not simply about checking for disasters; mostly it is checking on the mundane. With enStratus, I put a Python service on each server that checks for a variety of server health indicators—mostly related to capacity management. The service will notify the monitoring system if there is a problem with the server or its configuration and allow the monitoring system to take appropriate action. It also checks for the health of the applications running on the instance.

Load Balancer Recovery

One of the reasons companies pay absurd amounts of money for physical load balancers is to greatly reduce the likelihood of load balancer failure. With cloud vendors such as GoGrid—and in the future, Amazon—you can realize the benefits of hardware load balancers without incurring the costs. Under the current AWS offering, you have to use less-reliable EC2 instances. Recovering a load balancer in the cloud, however, is lightning fast. As a result, the downside of a failure in your cloud-based load balancer is minor.

Recovering a load balancer is simply a matter of launching a new load balancer instance from the AMI and notifying it of the IP addresses of its application servers. You can further reduce any downtime by keeping a load balancer running in an alternative availability zone and then remapping your static IP address upon the failure of the main load balancer.

Application Server Recovery

If you are operating multiple application servers in multiple availability zones, your system as a whole will survive the failure of any one instance—or even an entire availability zone. You will still need to recover that server so that future failures don't affect your infrastructure.

The recovery of a failed application server is only slightly more complex than the recovery of a failed load balancer. Like the failed load balancer, you start up a new instance from the application server machine image. You then pass it configuration information, including where the database is. Once the server is operational, you must notify the load balancer of the existence of the new server (as well as deactivate its knowledge of the old one) so that the new server enters the load-balancing rotation.

Database Recovery

Database recovery is the hardest part of disaster recovery in the cloud. Your disaster recovery algorithm has to identify where an uncorrupted copy of the database exists. This process may involve promoting slaves into masters, rearranging your backup management, and reconfiguring application servers.

The best solution is a clustered database that can survive the loss of an individual database server without the need to execute a complex recovery procedure. Absent clustering, the best recovery plan is one that simply launches a new database instance and mounts the still-functional EC2 volume formerly in use by the failed instance. When an instance goes down, however, any number of related issues may also have an impact on that strategy:

- The database could be irreparably corrupted by whatever caused the instance to crash.
- The volume could have gone down with the instance.
- The instance's availability zone (and thus the volume as well) could be unavailable.
- You could find yourself unable to launch new instances in the volume's availability zone.

On the face of it, it might seem that the likelihood of both things going wrong is small, but it happens. As a result, you need a fallback plan for your recovery plan. The following process will typically cover all levels of database failure:

1. Launch a replacement instance in the old instance's availability zone and mount its old volume.
2. If the launch fails but the volume is still running, snapshot the volume and launch a new instance in any zone, and then create a volume in that zone based on the snapshot.
3. If the volume from step 1 or the snapshot from step 2 are corrupt, you need to fall back to the replication slave and promote it to database master.
4. If the database slave is not running or is somehow corrupted, the next step is to launch a replacement volume from the most recent database snapshot.
5. If the snapshot is corrupt, go further back in time until you find a backup that is not corrupt.

Step 4 typically represents your worst-case scenario. If you get to 5, there is something wrong with the way you are doing backups.

HOW DO I KNOW MY BACKUPS ARE WORKING?

You need to test your disaster recovery procedures to know for sure that your backups are solid and the processes you have in place will work. I recommend testing your processes and your backups once every quarter, and certainly no less than once each year.

The most significant area in which your disaster recovery process can go wrong is in the database backups. Specifically, if you are not backing up your databases properly, the backups can appear to succeed but instead result in corrupted database backups. Silent failures are a bad thing.

The simplest way to test your database backup tools is to set up some bots to place a very high write transaction load on your database. Execute the backup script while the bots are running, and then try to recover the database from the backup. If your backup process is questionable, this test should leave you with a corrupted database.

CHAPTER SEVEN

Scaling a Cloud Infrastructure

One of the most useful features of cloud infrastructures is the ability to automatically scale an infrastructure vertically and horizontally with little or no impact to the applications running in that infrastructure. In truth, *useful* is an understatement. This feature fundamentally alters IT managers' relationships to their infrastructures and changes the way finance managers look at IT funding. But the feature is a double-edged sword.

The obvious benefit of cloud scaling is that you pay only for the resources you use. The noncloud approach is to buy infrastructure for peak capacity, waste resources, and pray your capacity planning was spot on. The downside of cloud scaling, however, is that it can become a crutch that lazy system architects use to avoid capacity planning. In addition, over-reliance on cloud scaling can lead an organization to respond to demand—and thus add cloud instances—when the demand in question simply has no business benefit.

In this chapter, I guide you through the issues of scaling a cloud infrastructure so that you can intelligently apply the full capabilities of the cloud to your IT infrastructure without falling prey to its dangers. Your success will start with reasonable capacity planning.

Capacity Planning

Capacity planning is basically developing a strategy that guarantees your infrastructure *can support* the resource demands placed on it. Covering the intricacies of capacity planning could fill an entire book. In fact, John Allspaw has done just that with his O'Reilly book titled *The Art of Capacity Planning (http://oreilly.com/catalog/9780596518578/index.html)*. I recommend reading that book, as capacity planning is truly important to success in the cloud.

For the purposes of this book, however, we will look at the core concerns for scaling in the cloud:

- Knowing your expected usage patterns as they vary during the day, over the course of a week, during holidays, and across the seasonal variance of your business
- Knowing how your application responds to load so that you can identify when and what kind of additional capacity you will need
- Knowing the value of your systems to the business so you can know when adding capacity provides value—and when it doesn't

Some look at the ability of cloud environments to automatically scale based on demand and think they no longer need to engage in capacity planning. Some look at capacity planning and think of tens or hundreds of thousands of dollars in consulting fees. Both thoughts are dangerous myths that must be put to rest.

Capacity planning is just as important in the cloud as it is in a physical infrastructure. And you do not need to engage in some outlandish capacity planning project to develop a proper plan. In the end, your objective is simply to make sure that when you incur additional cost by scaling your infrastructure, the additional cost will be supporting your objectives for that infrastructure.

Expected Demand

You absolutely need to know what demands you *expect* to be placed on your application. I am not suggesting you need to be a seer and accurately predict the number of page views on your website every day. You simply need to have a well-quantified expectation that will enable you to:

- Plan out an infrastructure to support expected loads
- Recognize when actual load is diverging in a meaningful way from expected load
- Understand the impact of changing application requirements on your infrastructure

The most obvious value of demand estimation is that—combined with understanding how your system responds to load—it tells you how many servers you need, what kind of servers you need, and what resources those servers require. If you have no idea how many people will use your website or web application, you literally have no idea whether the infrastructure you have put together will work right. It could fail within an hour of deployment, or you could waste countless amounts of money on unnecessary infrastructure.

You cannot possibly be expected to predict the future. The point of capacity planning is not to eliminate unexpected peaks in demand; you will always have unexpected peaks in demand. The point of capacity planning is to help you plan for the expected, *recognize* the unexpected, and react appropriately to the deviation.

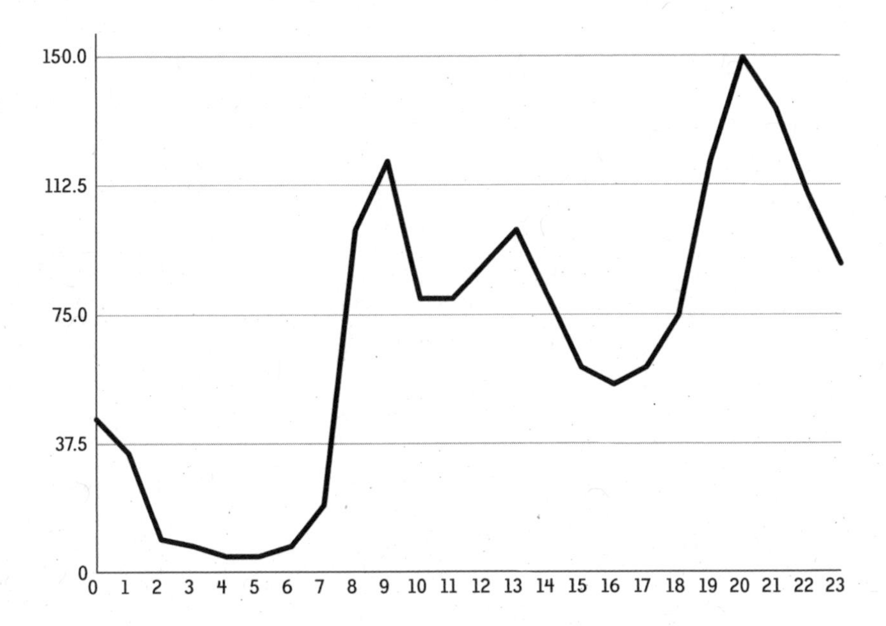

FIGURE 7-1. The projected daily load on an e-commerce site

Consider, for example, the scenario in which you have an infrastructure that supports 10 million transactions/second and you have a growth from your average load of 1 million transactions/second to 5 million transactions/second. If you had properly estimated the load, you would recognize whether the sudden surge was expected (and thus nothing to be concerned about) or unexpected and thus something to watch for potential capacity problems. Without proper load estimation, you would not know what to do about the variation.

Determining your expected demand

Figures 7-1 and 7-2 provide charts illustrating the expected traffic for an e-commerce site over the course of a typical day as well as projected peak volumes over the next 12 months.

The daily chart shows peaks in the morning, at lunch, and in the early evening. A major lull—almost nonexistent traffic—balances out these peaks in the early morning hours.

As a growing company, we expect gradually increasing volumes over the course of the year, punctuated by two product launches in May and September. Finally, we also have a seasonal increase in purchasing at the end of the year.

How do we get these numbers? As with many things, it depends on the application. In the case of the daily chart, historical patterns should form the basis of your expectations. If you have a steady business, the historical patterns are your expectations. A growing business, however, will see behavior alter with the changing market.

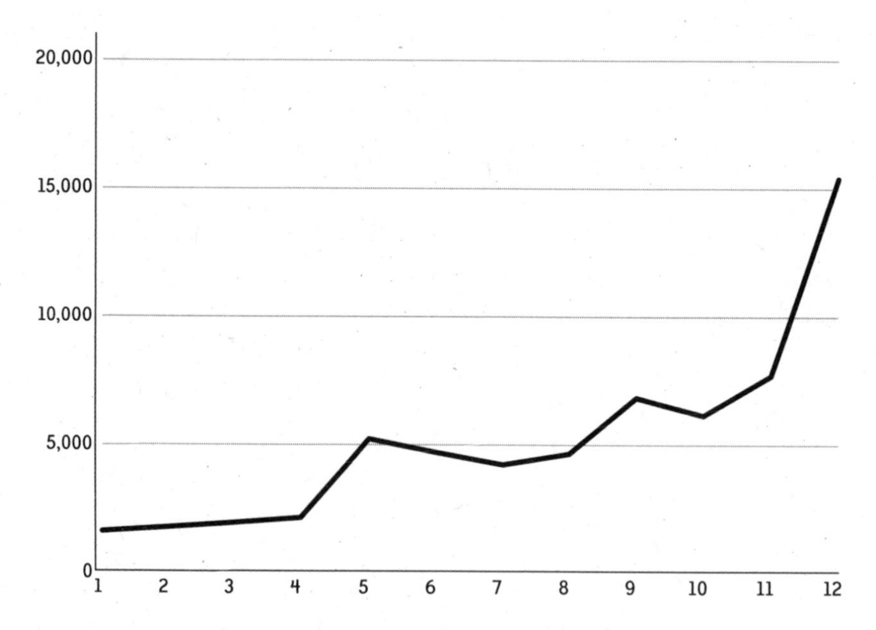

FIGURE 7-2. The expected load on the e-commerce site over the next 12 months

A more challenging situation is one in which you have no historical data. If you understand your market, a best guess is good enough until you develop historical data.

At the annual level, however, much of your projections are based on projections coming from other parts of the business. If you are selling goods to consumers, chances are that you expect seasonal variation, especially at the end of the year. Beyond general seasonal availability, however, you are at the mercy of what the business is doing to drive traffic to your website. You should therefore become good friends with sales, marketing, and anyone else who is driving traffic into the site. As we saw in Figure 7-2, you need to know about the intended product launch in May and the demand the company is projecting for the product in order to take that into account in your projections.

Analyzing the unexpected

You will see unexpected traffic. Sometimes the company launches a product that exceeds expectations; sometimes you receive unplanned media coverage; sometimes you are simply going to be wrong. Any time significant unexpected traffic hits your web systems, it's important to understand why traffic is varying from the unexpected. As we will discuss later in this section, that unexpected traffic could be either good news or bad news. It may require a change in all your projections, or it may simply be a wayward blip on the radar screen.

The Impact of Load

The ability to scale a web application or website is tied directly to understanding where the resource constraints lie and what impact the addition of various resources has on the application. Unfortunately, architects more often than not assume that simply adding another server into the mix can fix any performance problem. In reality, adding an application server into an infrastructure with a disk I/O bound database server will only make your problem worse. System architects must therefore understand the expected usage patterns of the applications they manage and execute tests to see how different scenarios create stress points in the infrastructure.

Assuming a perfectly efficient web application and database engine, the typical web application deployed into the cloud has all of these potential capacity constraints:

- The bandwidth into the load balancer
- The CPU and RAM of the load balancer
- The ability of the load balancer to properly spread load across the application servers
- The bandwidth between the load balancer and the application servers
- The CPU and RAM of the application server
- The disk I/O for read operations on the application server
- The write I/O for disk operations on the application server, a secondary disk constraint if the application is caching on the disk
- The bandwidth between the application server and any network storage devices (such as Amazon elastic block storage devices)
- The bandwidth between the application server and the database server
- The disk I/O for read operations on the read database
- The disk I/O for write operations on the write database
- The amount of space on the disk to support data storage needs

That's a lot of points where an application can run out of capacity. In reality, however, most of those are not the real problem. Your web application and database architectures are most likely to be the real problems.

Application architecture and database architecture revisited

In Chapter 4, I covered the proper way to architect an application for scaling in the cloud. To summarize, you follow the same rules that enable an application to scale horizontally outside of the cloud. You should follow these guidelines:

- Use the fastest storage devices available to you for database access
- Avoid keeping transactional data at the application server layer

- Enable multiple copies of your application server to run against the same database without any communication between the application servers
- Properly index your database
- If possible, use a master/slave setup with read operations directed to slaves
- With Amazon EC2, design your redundancies to minimize traffic across availability zones

Points of scale

Depending on your application, the most likely initial stress points will be one of the following three components:

- The CPU on your application server
- The RAM on your application server
- The disk I/O on your database server

Every application has stress points. If it didn't, it could run on a single Intel 386 under indefinite load. For the Valtira application (a Java-based web application I architected for the company by the same name), our first bottleneck is always CPU on the application server. As it scales horizontally, CPU ceases to be a significant factor, but (depending on the content inside the Valtira deployment) we encounter a new bottleneck on either bandwidth into the network or database disk I/O. In the cloud, it's the disk I/O.

So our next point of scale is to split out read operations across multiple database slaves and use the master only for write operations.* The next choke point tends to become the disk I/O for write operations on the database master. At that point, we need to segment the database or look at a more expensive database solution.

In other words, we have a firm grasp over how our application operates, where we run into capacity constraints, and what we can do about those capacity constraints. Without that knowledge, I might be deluded into believing the solution is always to add application servers.

Of course, you could do a very expensive analysis to try to determine the exact number of users at which you need to take each scaling action. That's not really necessary, however. Just launch your environment at the expected level of scale, begin running realistic load tests, and make notes on the impact of load and additional scale. It should take very little time and cost very little money, and the information you use should be good enough, except for situations in which hiring expert capacity planners is absolutely necessary.

* To be able to split read operations from write operations, you need to have some protection in your application against dirty writes, as described in Chapter 4.

The Value of Your Capacity

In a web application, you don't simply add more capacity to the system just because your CPUs have hit 90% utilization. It's important to be able to answer the question, "What does supporting this additional load get me?" Knowing the value of the demand on your system will help answer that question.

In a grid computing system, it's often easy to understand the value of additional load. For example, a video-rendering farm scales specifically to render more video. If that's your business, you should be able to determine what rendering each video is worth to you and do an appropriate cost/benefit analysis. Understanding whether you are launching more capacity to support one video—as opposed to waiting for existing capacity to become available—will help you be more efficient in your spending. For the most part, however, the decision to add capacity in many nonweb systems is generally straightforward.

Web applications aren't that cut-and-dried. A website or web application typically supports dozens or hundreds of different use cases, each with its own value to the business. The value on an e-commerce site of the shopping experience is much different from the value of the CMO's blog. A 100% spike in activity related to a successful product promotion is thus more important than a 100% spike in activity due to a Twitter reference to the CMO's latest blog entry.

A simple thought experiment

Let's look at a simple example of what I am talking about. We have a simple web application with basic corporate content that serves as our main sales generation tool. We use SalesForce.com's Web2Lead to take leads from the website and get them into SalesForce.com. An intranet component behind the website enables the marketing team to set up campaigns and landing pages and do basic reporting.

The website begins seeing a sudden, unexpected spike in activity. As load approaches capacity, do we add more capacity or do we let it ride the storm?

To make the call, we need to know how much it is going to cost to add capacity and the value of that capacity to the business. The critical questions we need to answer are:

- How is the lack of capacity impacting site visitors?
- Do we think this spike represents appropriate uses of the system? Is there a bigger issue we need to be concerned with?
- Do we expect the demand to increase further? Do we think we are near the peak?
- Does it cost us anything to let things ride? If so, is that cost greater than the cost of adding capacity into the system?

For the purposes of this thought experiment, we will assume all scaling is done manually. Now we can do some research. Imagine we find the following answers to our questions:

How is the lack of capacity impacting site visitors?
: As we approach capacity, the website begins to slow down and the system stops processing logins. If the system gets too bogged down, the speed of the site will eventually become unusable.

Do we think the spike is legitimate?
: An analysis of site traffic shows that the CMO said something controversial in her blog and a number of social media sites picked up on it. The blog is being hammered pretty heavily, but the rest of site traffic is pretty much normal.

Do we expect things to get worse?
: It does not look like it, as the traffic has begun to level out. But is this a new usage plateau?

Does it cost us anything to let things ride?
: The cost of letting this play out is largely insignificant, as regular site visitors are still getting through. The only people significantly impacted are internal marketing users who can put up with a few hours of limited or no access. On the other hand, because we are in the cloud and we have done load testing, we know that adding a single additional application server will return system operation to normal. In other words, it will cost us a few dollars.

The answer
: We add the capacity. It does not buy us much, but it also does not cost us much.

How might the outcome have been different?

The most important lesson of this experiment is how different the results and our decision would have been if we were not in a cloud. The cost of adding more capacity would have been huge and, by the time we got it, the unexpected demand would have subsided.

Another outcome worth considering, however, is if there were something peculiar about the traffic—for instance, if it appeared to be some kind of out-of-control botnet looking for site vulnerabilities. The issues become much more difficult in that scenario because it is unclear whether adding capacity is going to help you. In fact, you might end up in a situation in which as you add capacity, the botnet increases its probes ands forces you to add more capacity. In short, you would be in a spiral in which you increasingly become a victim to the botnet. The net impact of additional capacity on normal traffic is negligible. As a result, you could be incurring greater cost and greater exposure to an external threat without incurring any business benefit.

The key point here is that additional demand does not automatically mean that you must add more capacity.

Cloud Scale

The cloud empowers you to alter your computing resources to meet your load requirements. You can alter your capacity both manually (by executing a command on a command line or through a web interface) and programmatically (through predefined changes in capacity or through software that automatically adjusts capacity to meet actual demand).

The ability to manually adjust capacity is a huge advantage over traditional computing. But the real power of scaling in the cloud lies in dynamic scaling.

Dynamic scaling
: This term—which I sometimes also refer to as cloud scaling—enables software to adjust the resources in your infrastructure without your interactive involvement. Dynamic scaling can take the form of *proactive scaling* or *reactive scaling*.

Proactive scaling
: This involves a schedule for altering your infrastructure based on projected demand. If you consider the application described back in Figure 7-1, we would configure our cloud management tools to run with a minimal infrastructure that supports our availability requirements during the early morning hours, add capacity in the late morning, drop back to the baseline until lunch, and so on. This strategy does not wait for demand to increase, but instead increases capacity based on a plan.

Reactive scaling
: In this strategy, your infrastructure reacts to changes in demand by adding and removing capacity on its own accord. In the capacity valuation thought experiment, an environment engaging in reactive scaling might have automatically added capacity when it detected the unexpected spike in activity on the CMO blog.

> **NOTE**
>
> **If you have read my blog online, you may have noticed that in the past I have used the terms "dynamic scaling" to refer to proactive scaling and "auto-scaling" to refer to reactive scaling. I have altered the terminology here because "dynamic" and "automatic" have proven to be confusing terms, whereas the difference between proactive and reactive is fairly easy to understand.**

Tools and Monitoring Systems

Throughout this book, I have referred to cloud infrastructure management tools and monitoring systems as being critical to the management of a cloud infrastructure. I run one such company, enStratus, but there are a number of other good systems out there, including RightScale and Morph. Which one is right for you depends on your budget, the kinds of applications you manage, and what parts of infrastructure management matter most to you.

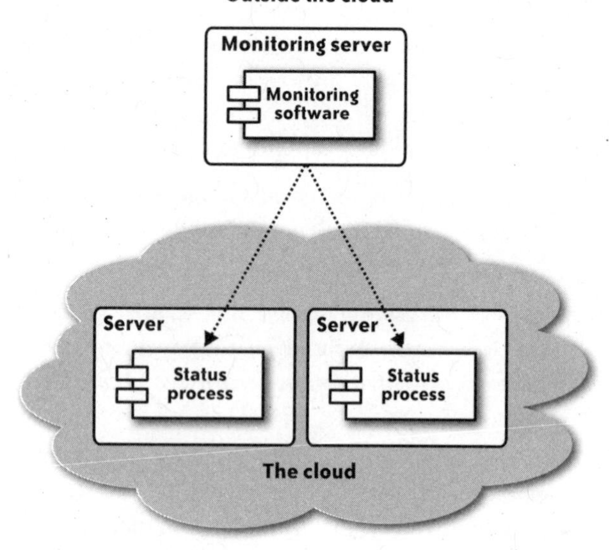

FIGURE 7-3. General architecture for a system monitoring your cloud health

Whatever tool you pick, it should minimally have the following capabilities (and all three services I mentioned have them):

- To schedule changes in capacity for your application deployments
- To monitor the deployments for excess (and less than normal) demand
- To adjust capacity automatically based on unexpected spikes or falloffs in demand

Monitoring involves a lot more than watching for capacity caps and switching servers off and on. In Chapter 6, I covered the role monitoring plays in watching for failures in the cloud and recovering from those failures. A notification system is obviously a big part of monitoring for failures. You should have full logging of any change in capacity—whether scheduled or not—and email notifications of anything extraordinary.

You can also roll your own monitoring system if you don't want to pay for someone else's software. A monitoring system has the architecture described in Figure 7-3.

Compared to disaster recovery, it's not as critical for capacity planning purposes that your monitoring server be outside the cloud. Nevertheless, it's a very good idea, and a critical choice if bandwidth management is part of your monitoring profile.

The monitoring checks on each individual server to get a picture of its current resource constraints. Each cloud instance has a process capable of taking the vitals of that instance and reporting back to the monitoring server. Most modern operating systems have the ability to operate as the status process for core system data, such as CPU, RAM, and other SNMP-related data. In addition, Java application servers support the JMX interfaces that enable you to query the performance of your Java virtual machines.

For security purposes, I prefer having the monitoring server poll the cloud instances rather than making the cloud instances regularly report into the monitoring server. By polling, you can put your monitoring server behind a firewall and allow no incoming traffic into the server. It's also important to keep in mind that you need the ability to scale the monitoring server as the number of nodes it must monitor grows.

The process that checks instance vitals must vary for each instance based on its function. A load balancer can be fairly dumb, and so all the monitor needs to worry about is the server's RAM and CPU utilization. A database server needs to be slightly more intelligent: the vitals process must review disk I/O performance for any signs of trouble. The most difficult monitoring process supports your application servers. It should be capable of reporting not simply how much the instance's resources are being utilized, but what the activity on the instance looks like.

The monitoring server then uses analytics to process all of that data. It knows when it is time to proactively add and remove scale, how to recognize unexpected activity, and how to trigger rules in response to unexpected activity.

The procurement process in the cloud

Whether you scale dynamically or through a human pulling the levers, the way you think about spending money in the cloud is very different from a traditional infrastructure. When you add new resources into your internal data center or with a managed services provider, you typically need to get a purchase order approved through your company procurement processes. Finance approves the purchase order against your department's budget, and the order goes off to the vendor. You don't spend $3,000 on a server unless that spend is cleared through Finance.

Nothing in the AWS infrastructure prevents you from executing *ec2-run-instances* just one time on an EC2 extra-large instance and spending $7,000 over the course of a year. Anyone who has access to launch new instances or alter the scaling criteria of your cloud management tools has full procurement rights in the cloud. There's no justification that an IT manager needs to make to Finance; it's just a configuration parameter in a program that Finance may never touch.

Finance should therefore be involved in approving the monthly budget for the team managing the cloud infrastructure. Furthermore, controls should be in place to make sure any alterations in the resources you have deployed into the cloud are aligned with the approved budget. If you don't put this human process in place, you may find Finance turning from the biggest supporter of your move into the cloud to your biggest critic.

Managing proactive scaling

A well-designed proactive scaling system enables you to schedule capacity changes that match your expected changes in application demand. When using proactive scaling, you should understand your expected standard deviation. You don't need to get out the statistics textbooks...or worse, throw out this book because I mentioned a term from statistics. I simply mean you should roughly understand how much normal traffic deviates from your expectations. If you expect site activity of 1,000 page views/hour around noon and you are seeing 1,100, is that really unexpected? Probably not.

Your capacity for any point in time should therefore be able to handle the high end of your expected capacity with some room to spare. The most efficient use of your resources is just shy of their capacity, but scheduling things that way can create problems when your expectations are wrong—even when combined with reactive scaling. Understanding what constitutes that "room to spare" is probably the hardest part of capacity planning.

Managing reactive scaling

Reactive scaling is a powerful rope you can easily hang yourself with. It enables you to react quickly to unexpected demand. If you fail to do any capacity planning and instead rely solely on reactive scaling to manage a web application, however, you probably will end up hanging yourself with this rope.

The crudest form of reactive scaling is utilization-based. In other words, when your CPU or RAM or other resource reaches a certain level of utilization, you add more of that resource into your environment. It makes for very simple logic for the monitoring system, but realistically, it's what you need only a fraction of the time. We've already seen some examples of where this will fail:

- Hiked-up application server processing that suffers from an I/O bound database server. The result is increased loads on the database server that further aggravate the situation.
- An attack that will use up whatever resources you throw at it. The result is a spiraling series of attempts to launch new resources while your costs go through the roof.
- An unexpected spike in web activity that begins taxing your infrastructure but only mildly impacts the end user experience. You know the activity will subside, but your monitor launches new instances simply because it perceives load.

A good monitoring system will provide tools that mitigate these potential problems with reactive scaling. I have never seen a system, however, that is perfectly capable of dealing with the last scenario. It calls for an understanding of the problem domain and the pattern of activity that determines you should not launch new instances—and I don't know of an algorithmic substitute for human intuition for these decisions.

However your monitoring system defines its rules for reactive scaling, you should always have a governor in place. A governor places limits on how many resources the monitoring system can automatically launch, and thus how much money your monitoring system can spend on your behalf. In the case of the attack on your infrastructure, your systems would eventually end up grinding to a halt as you hit your governor limit, but you would not end up spending money adding an insane number of instances into your cloud environment.

A final issue of concern that affects both proactive and reactive scaling—but more so for reactive scaling—is the fallibility of Amazon S3 and the AWS APIs. If you are counting on reactive scaling to make sure you have enough resources, Amazon S3 issues will weaken your plans. Your system will then fail you.

A recommended approach

I am not terribly fond of reactive scaling, but it does have its uses. I prefer to rely heavily on proactive scaling to manage my infrastructure based on expected demand, including excess capacity roughly one to two times the difference between expected demand and the highest expected demand (about three to five standard deviations from the expected demand). With this setup, reactive scaling should almost never kick in.

Unexpected demand does occur. Instead of using reactive scaling to manage unexpected load, I instead use it to give people time to react and assess the situation. My governors are thus generally set to 150% to 200% of the baseline configuration. In other words, if my baseline configuration has two application servers and my application will scale to 200% of the baseline capacity by adding two more application servers, I direct the governor to limit scaling at two application servers and notify me well in advance of the need to add even one.

As a result of using reactive scaling in this way, my infrastructure should scale in a controlled manner automatically in reaction to the unexpected demand up to a certain point. I should also have the time to examine the unexpected activity to determine whether I want to bump up the governors, change my baseline configuration, or ignore the event altogether. Because I am not running near capacity and I am not relying on reactive scaling to keep things running, Amazon S3 failures are unlikely to impact me.

Although the absolute numbers I have mentioned may not make sense for your web application, the general policy should serve you well. Whatever numbers make sense for your application, you should have an approved budget to operate at the peak capacity for at least long enough a time period to approve increasing the budget. If your budget is approved for expected capacity and you find yourself operating at the limits of your governors, your finance department will not be pleased with you.

Scaling Vertically

So far, I have been entirely focused on horizontal scaling, which is scaling through the addition of new servers. Vertical scalability, on the other hand, is scaling by replacing an existing server with a beefier one or a more specialized one. All virtualized environments—and cloud environments in particular—are very good at horizontal scaling. When it comes to vertical scaling, however, the cloud has some important strengths and some very important weaknesses.

The strength of the cloud with respect to vertical scaling is the ease with which you can try out smaller or less-specialized configurations and the ability to make the system prove the need for more configurations. Clouds (and the Amazon cloud more so than its competitors) do a poor job of providing specialized system configurations.

Amazon currently provides you with five different system choices. If a component of your application requires more RAM than one of the Amazon instances supports, you are out of luck. GoGrid, in contrast, provides a greater degree of customization, including the ability to design high I/O configurations. In the end, none of these options will match your ability to configure a very specialized system through Dell's configurator.

Though I have spent a lot of time in this book talking about horizontal scaling, I always scale vertically first.

START SMALL

When we build applications at Valtira, we start on Amazon's medium instances and force the application to prove it needs a large or extra-large instance. In the end, large instances cost four times as much as medium instances. If you can get the same performance spreading your application across four medium application servers as you get from one large instance, you are better off using the medium instance, since you get availability and performance this way. You get only performance with the large instance.

Vertical scalability in the Amazon cloud is most effective when you need more RAM. The Valtira application I mentioned earlier in this chapter is an excellent example of such an application. I left out that the first point of scale for Valtira is actually RAM. Most systems deployed on the Valtira platform don't need a lot of RAM—1 to 2 GB is generally sufficient. Some applications that leverage certain components of the Valtira platform, however, require a lot more RAM. Because Valtira will essentially duplicate its memory footprint across all servers in a cluster, adding more servers into the equation does not help at all. Moving to a server with more RAM, however, makes all the difference in the world.

Vertical scalability can help with other capacity constraints as well. Table 7-1 describes how a theoretical application responds to different kinds of scaling.

TABLE 7-1. Example of Amazon server CPU options

Configuration	Capacity	Cost
Eight Amazon medium	8,000 page views/minute	$0.80/hour
Two Amazon large	10,000 page views/minute	$0.80/hour
One Amazon extra-large	10,000 page views/minute	$0.80/hour

If you assume linear scalability horizontally, you want to switch the infrastructure from eight medium instances to two large instances rather than adding a ninth medium instance. Admittedly, this example is absurdly simplistic. The point, however, is that sometimes it simply makes financial sense to scale vertically.

Vertical dynamic scaling is trickier than horizontal. More to the point, scaling vertically is a special case of horizontal scaling. It requires the following delicate dance:

1. Add an instance into the cloud of the beefier system, as if you were scaling horizontally. The only difference is that you are using one of the larger machine instances instead of a duplicate of the existing infrastructure instances.
2. Wait for the new instance to begin responding to requests.
3. Remove one or more of the old, smaller instances from the system.

When you put horizontal scaling together with vertical scaling, you end up with an infrastructure that makes the most efficient use of computing resources.

APPENDIX A

Amazon Web Services Reference

AMAZON WEB SERVICES ARE ACCESSIBLE PRIMARILY through SOAP and REST web services with higher-level command lines and language-specific APIs built on top of them. This appendix describes the commands that wrap around those APIs.

Many different higher-level abstractions exist that implement the SOAP and APIs in different programming languages. If you are building tools to manage EC2 or S3, you should identify the library best suited to your programming language and functional needs.

Amazon EC2 Command-Line Reference

The Amazon EC2 command-line tools are wrappers around the web services API. In fact, there is basically a one-to-one mapping between a command line, its arguments, and an API call of a similar name with similar parameters.

Every command has the following general form:

```
command [GENERAL OPTIONS] [COMMAND OPTIONS]
```

For example, the command to start an EC2 instance looks like this:

```
ec2-run-instances -v ami-123456 -g dmz
```

In this instance, the `-v` is a general option for specifying verbose output, and `ami-1234566 -g dmz` are the command-specific options.

The general options are:

`-`
: Pull in command parameters from standard input.

`-C` *`certificate`*
: The certificate to authenticate your web services request with Amazon. This value overrides the environment variable `EC2_CERT`.

`--connection-timeout`
: Indicates an alternate SOAP connection timeout in seconds.

`--debug`
: Prints out debug information.

`--headers`
: Display column headers.

`--help`
: Prints out help for the command in question.

`-K` *`privatekey`*
: The private key to authenticate your web services request with Amazon. This value overrides the environment variable `EC2_PRIVATE_KEY`.

`--region` *`region`*
: Specifies the region to apply to the command.

`--request-timeout`
: Indicates an alternate SOAP request timeout in seconds.

`--show-empty-fields`
: Displays empty columns in command responses as `(nil)`.

`-U` *`url`*
: Specifies the Amazon Web Services URL to make the API call against. This option overrides the `EC2_URL` environment variable.

`-v`
: Indicates that you would like verbose output. Verbose output shows the SOAP requests and responses.

ec2-add-group

`ec2-add-group` *`groupname`* `-d` *`description`*

Adds a new security group into your Amazon EC2 environment. The group name you provide will identify the group for use in other commands. The description is simply a user-friendly description to help you remember the purpose of the group.

The new group allows no external access to instances launched in it. You must call *ec2-authorize* before any traffic will route to instances in this group.

Example

```
$ ec2-add-group mydmz -d DMZ
GROUP mydmz DMZ
```

ec2-add-keypair

`ec2-add-keypair` *`keyname`*

Creates a new RSA 2048-bit encryption key pair. The name you specify in creating the key will be used for referencing the key using other commands or the EC2 web services API. The command stores the public key with Amazon and displays the private key to stdout. You should then securely store the private for use in accessing instances you launch using this key pair.

Example

```
$ ec2-add-keypair georgekey
KEYPAIR georgekey  2e:82:bb:91:ca:51:22:e1:1a:84:c8:19:db:7c:8b:ad:f9:5e:27:3e
-----BEGIN RSA PRIVATE KEY-----
MIIEowIBAAKCAQEAkot9/O9tIy5FJShOX5vLGCu3so5Q4qG7cU/MBm45c4EVFtMDpU1VpAQi1vn9
r7hr5kLr+ido1d1eBmCkRkHuyhfviJmH1FTOWm6JBhfOsOgDUOpInyQOPOnRFLx4eyJfYsiK/mUm
hiYC9Q6VnePjMUiHSahOL95C8ndAFBlUAuMDDrXMhLypOGRuWkJo+xtlVdisKjlOTOl33q3VSeT6
NBmZwymWOguGWgKWMpzpDLhV9jhDhZgaZmGUKPOwPQqdV6psA9PuStN1LJkhWVYuQTqH9UUolvJn
ZXx5yE2CSpPW+8zMb4/xUuweBQ6grw8O3IxhKWbFCpGGhkpk5BB+MQIDAQABAoIBAQCIs6U6mA4X
5l7MFdvRIFSpXIbFAutDLlnbjvOlAAeJztOsaHWbKvP7x3vOjElxNRk6OC1HMqIh9plyW46Cl5i4
XvGsvIOvt9izFS+vRmAiOJx5gu8RvSGpOiPXMyUOwFC4ppi6TQNN2oGhthQtsFrMK3tAY8dj8fMD
mehll2b+NPZRWPp9frm3QtwLIOMeWm1ntknCVSjBqj21XRg3UPbE8r8ISlSGryqJBAOKjnOj+cMN
2SBx8iC+BHxD9xSUvXs4hVjUpQofzd+8BAZbsXswj+/ybuq1GlNwzpUKKEfH1rN3TZztywN5Z9Hb
EbkOtgRYi/2htSpbuDq5b/cTaxIRAoGBAOLRgfZhEwnGQvveMOhRLLko1D8kGVHP6KCwlYiNowO7
G8FkP6U3wcUrsCTtvOFB/79FeWVT+o7v25D34cYFtGbfnp3Zh9bdTBi18PbIHQHvD4tIAIF+4PcO
XMRsJCrzhChOLY1G/laMi5EKFcx6RU8Pjup92YbEbi/fkybcrmS9AoGBAKVmGI5PVOOA1OLQkTov
CnLuyfAPL9s8w6eOy+9WMRd8+27tI6H8yGuEdsF9X9bOJQsnTM3+A+RC8ylVXWPgiflbcpbPsZ8a
HVuTg37D/iFpl42RzrMtzhgLCahvNotirNyAYnklBsOlmtsQdJSJOGPpv4SOloSoPT+jbP4ONUiF
AoGAWU48aZHXOSYDAcB+aTps7YqR5zqDbZ767SoZ9mYuKOt5BjA+jwLhHIoTEbc5gOfFNr5YCfmC
OfzG6tFu59UfLtIlVelsfsErUR9x/PjVOwkZibGT4Wjfkubox738j5zKEESXOuR9B/7WhQj/hD8w
QuzRTKq4lOOITvksqOSAtdECgYAqpr1GVWdpOAGylR4eJutG4BTq9r+chXrexpAIU+2s5OnhnP1H
VGxKbYpCMxZ3ygj7a1L++7X9MtaJnh3LF6f8yXwvL7faE13ms4+BLQFnlFckhqkKw5EV2iLPcH5c
SOHQSrsaClZINXhNbVziwPcgDLL6d9qQsuG4e2gry3YqEQKBgFHqE4UJCOd5WiAGONOcYDTF/wh6
iujW5tY9OF63xAn2B236DGE+8o2wGwU77u59LO7jyx4WyR8TpcorL79zZuzmOVjn9nslAu7tkS6O
wmdEMOO2LrGnKGydSdRF5ONH8TgbO6Otxh+hWyWtvkPOtSOZyGu1z7S7JSOZPX42Arm8
-----END RSA PRIVATE KEY-----
```

ec2-allocate-address

`ec2-allocate-address`

Allocates a new Amazon elastic IP address and prints out the allocated address. This newly allocated address then becomes available only to you, to assign to instances as you wish. Amazon charges for the nonuse of elastic IP addresses as well as excessive reassignment of addresses.

Example

```
$ ec2-allocate-address
ADDRESS 67.202.55.255
```

ec2-associate-address

`ec2-associate-address -i instanceid ipaddress`

Associates an address obtained through `ec2-allocate-address` with a running EC2 instance. This command will also disassociate the IP address with any current associations it might have. It is therefore important you not accidentally use this method to reassign an IP unless you intend the reassignment.

Example

```
$ ec2-associate-address -i i-12b3ff6a 67.202.55.255
ADDRESS 67.202.55.255    i-12b3ff6a
```

ec2-attach-volume

`ec2-attach-volume volumeid -i instanceid -d device`

Attaches an existing elastic block storage volume to a running EC2 instance, exposing it to the instance as the specified device. The proper device name is platform-dependent, with Linux variants expecting device names such as */dev/sdh* and Windows expecting device names such as *xvdh*.

With the block storage volume attached to your instance, you can mount and format it from the instance using the operating-system-specific disk management utilities.

The command output describes the state of the attaching volume with its attachment information.

Example

```
$ ec2-attach-volume vol-81aeb37f -i i-12b3ff6a -d /dev/sdf
ATTACHMENT vol-81aeb37f i-12b3ff6a /dev/sdf attaching 2008-12-17T22:36:00+0000
```

ec2-authorize

```
ec2-authorize groupname -P protocol (-p portrange
 | -t icmptypecode) [-u sourceuser ...] [-o sourcegroup ...]
 [-s sourceaddress]
```

Authorizes network traffic to EC2 instances launched with the specified group name.

You can authorize incoming traffic based on a variety of criteria:

- Based on the subnet from which the traffic is originating
- Based on the group membership of the EC2 instance from which the traffic is originating (if originating from an EC2 instance)

- Based on the protocol (TCP, UDP, ICMP) of the traffic
- Based on the destination port of the traffic

By default, a group allows no traffic to pass to EC2 instances that belong to it (though members may belong to multiple groups, in which case their membership in other groups may allow it). To get traffic flowing, you must specifically authorize traffic to flow.

Except when you want to enable traffic to flow from one EC2 group to another, you can control traffic down to the protocol and port level. If you allow traffic to flow from one group to another, it's an all-or-nothing proposition.

Examples

```
# Grant port 80 access to all traffic regardless of source
$ ec2-authorize mydmz -P tcp -p 80 -s 0.0.0.0/0
GROUP          mydmz
PERMISSION     mydmz    ALLOWS    tcp    80    80    FROM    CIDR    0.0.0.0/0

# Grant access to the app server group from the DMZ group
$ ec2-authorize myapp -u 999999999999 -o mydmz
GROUP          myapp
PERMISSION     myapp    ALLOWS    all    FROM    USER    999999999999 GRPNAME mydmz

# Grant access to a range of ports from a specific IP address
$ ec2-authorize mydmz -P udp -p 3000-4000 -s 67.202.55.255/32
GROUP          mydmz
PERMISSION     mydmz    ALLOWS    udp    3000    4000    FROM    CIDR 67.202.55.255/32
```

ec2-bundle-instance

`ec2-bundle-instance` *`instanceid`* `-b` *`s3bucket`* `-p` *`prefix`* `-o` *`accesskey`* `(-c` *`policy`* `| -w` *`secretkey`*`)`

Windows instances only. Bundles your Windows instance and stores it in Amazon S3 to be registered using *ec2-register.*

Example

```
$ ec2-bundle-instance i-12b3ff6a -b mybucket -p myami -o 999999999999 -w
1Y1zp/1iKzSAg9B041Q0T3gMxje7IfnXtN5asrM/dy==
BUNDLE  bun-abd5209d8  i-12b3ff6a  mybucket myami  pending  2008-12-
18T13:08:18+0000  2008-12-18T13:08:18+0000
```

ec2-cancel-bundle-task

`ec2-cancel-bundle-task` *`bundleid`*

Windows instances only. Cancels a bundle process currently underway.

Example

```
$ ec2-cancel-bundle-task bun-abd5209d8
BUNDLE bun-abd5209d8 i-12b3ff6a  mybucket  myami  canceling  2008-12-
18T13:13:29+0000  2008-23-18T13:13:29+0000
```

ec2-confirm-product-instance

ec2-confirm-product-instance *productcode* -i *instanceid*

Enables an AMI owner to check whether the specified instance has the named product code attached to the instance.

Example

```
$ ec2-confirm-product-instance zt1 -i i-12b3ff6a
ztl i-12b3ff6a false
```

ec2-create-snapshot

ec2-create-snapshot *volumeid*

Creates a differential snapshot of the specified volume ID and stores the snapshot in Amazon S3. It is best to take snapshots of "frozen" filesystems to avoid data integrity concerns. It is OK to begin writing to the volume again after this command has returned successfully.

Example

```
$ ec2-create-snapshot vol-12345678
SNAPSHOT snap-a5d8ef77  vol-12345678  pending  2008-12-20T20:47:23+0000
```

ec2-create-volume

ec2-create-volume (-s *size* | --snapshot *snapshotid*) -z *zone*

Creates a new volume either with the specified volume size or based on the specified snapshot in the named availability zone. You must specify an availability zone and either a size or a snapshot. The size parameter is the size in gigabytes.

Examples

```
# Create a new volume of 10 GB
$ ec2-create-volume -s 10 -z eu-west-1a
VOLUME  vol-12345678  10  eu-west-1a  creating  2008-12-20T20:47:23+0000

# Create a volume based on a stored snapshot
$ ec2-create-volume --snapshot snap-a5d8ef77 -z eu-west-1a
VOLUME  vol-12345678  10  eu-west-1a  creating  2008-12-20T20:47:23+0000
```

ec2-delete-group

`ec2-delete-group` *`group`*

Deletes the specified security group from your account. You cannot delete a group until all references to it have been removed (instances in the group and other groups with rules allowing access to this group).

Example

```
$ ec2-delete-group mydmz
GROUP mydmz
```

ec2-delete-keypair

`ec2-delete-keypair` *`keypair`*

Deletes the specified public key associated with the named key pair from your Amazon account.

Example

```
$ ec2-delete-keypair georgekey
KEYPAIR georgekey
```

ec2-delete-snapshot

`ec2-delete-snapshot` *`snapshotid`*

Deletes a snapshot from your account.

Example

```
$ ec2-delete-snapshot snap-a5d8ef77
SNAPSHOT snap-a5d8ef77
```

ec2-delete-volume

`ec2-delete-volume` *`volumeid`*

Deletes a volume from your account.

Example

```
$ ec2-delete-volume vol-12345678
VOLUME vol-12345678
```

ec2-deregister

`ec2-deregister` *`imageid`*

Deregisters a machine image so that you can no longer launch instances from it. You must delete the AMI from S3 separately in order to free up any space associated with the AMI.

Example

```
$ ec2-deregister ami-f822a39b
IMAGE ami-f822a39b
```

ec2-describe-addresses

`ec2-describe-addresses [ipaddres1 [...ipaddressN]]`

Lists the information associated with the specified elastic IP addresses. If you do not specify any particular addresses, it will list all IP addresses you have allocated in the current EC2 region.

Examples

```
# SHOW ALL ALLOCATED
$ ec2-describe-addresses
ADDRESS 67.202.55.255 i-12b3ff6a
ADDRESS 67.203.55.255

# SHOW A SPECIFIC ADDRESS
$ ec2-describe-addresses 67.202.55.255
ADDRESS 67.202.55.255 i-12b3ff6a
```

ec2-describe-availability-zones

`ec2-describe-availability-zones [zone1 [...zoneN]]`

Lists the information associated with the specified EC2 availability zones. If no zone is specified, it will list all zones associated with the current EC2 region.

Examples

```
# SHOW ALL ALLOCATED
$ ec2-describe-availability-zones
AVAILABILITYZONE    us-east-1a  available
AVAILABILITYZONE    us-east-1b  available
AVAILABILITYZONE    us-east-1c  available

# SHOW A SPECIFIC ZONE
$ ec2-describe-availabilty-zones us-east-1a
AVAILABILITYZONE    us-east-1a  available

# SHOW ALL EU ZONES
$ ec2-describe-availability-zones --region eu-west-1
AVAILABILITYZONE    eu-west-1a  available
AVAILABILITYZONE    eu-west-1b  available
```

ec2-describe-bundle-tasks

ec2-describe-bundle-tasks [*bundle1* [...*bundleN*]]

For Windows instances only. Lists the information associated with the specified bundle tasks. If no specific task is named, the command will list all bundle tasks associated with this account.

Examples

```
# SHOW ALL TASKS
$ ec2-describe-bundle-tasks
BUNDLE  bun-abd5209d8  i-12b3ff6a  mybucket myami  pending  2008-12-
18T13:08:18+0000  2008-12-18T13:08:18+0000
BUNDLE  bun-abd5209d9  i-12b3ff7a  mybucket myami  pending  2008-12-
18T13:08:18+0000  2008-12-18T13:08:18+0000

# SHOW SPECIFIC TASK
$ ec2-describe-bundle-tasks bun-abd5209d8
BUNDLE  bun-abd5209d8  i-12b3ff6a  mybucket myami  pending  2008-12-
18T13:08:18+0000  2008-12-18T13:08:18+0000
```

ec2-describe-group

ec2-describe-group [*group1* [...*groupN*]]

Lists the information associated with the specified security groups. If no groups are specified, then all groups associated with the account will be listed.

Examples

```
# SHOW ALL GROUPS
$ ec2-describe-group
GROUP mydmz DMZ
PERMISSION    mydmz    ALLOWS    tcp    80    80    FROM    CIDR    0.0.0.0/0
GROUP myapp App
PERMISSION    myapp    ALLOWS    all    FROM    USER    999999999999 GRPNAME mydmz

# SHOW A SPECIFIC GROUP
$ ec2-describe-group mydmz
PERMISSION    mydmz    ALLOWS    tcp    80    80    FROM    CIDR    0.0.0.0/0
```

ec2-describe-image-attribute

ec2-describe-image-attribute *imageid* (-l | -p)

Describes the attributes for a specific AMI. You specify whether you want to see the launch permissions or the product codes.

Examples

```
# SHOW LAUNCH PERMISSIONS
$ ec2-describe-image-attribute ami-f822a39b -l
launchPermission  ami-f822a39b  userId 999999999999
```

```
# SHOW PRODUCT CODE
$ ec2-describe-image-attribute ami-f822a39b -p
productCodes ami-f822a39b productCode zz95xy
```

ec2-describe-images

`ec2-describe-images [imageid1 [...imageidN]] [-a] [-o ownerid] [-x ownerid]`

Describes the information associated with a specific image ID or any images that match the specified parameters. If you specify no parameters at all, the command will list out all images owned by your account.

Specific options include:

`-a`
: Lists AMIs for which the user has execution rights.

`-o` *`ownerid`*
: Lists AMIs belonging to the specified owner or owners. You may also use the special owner IDs: `amazon` (for public images), `self` (referring to your own images), and `explicit` (referring to the images for which you have launch permissions).

`-x` *`ownerid`*
: Lists AMIs for which the specified owner or owners have launch permissions. In addition to a standard owner ID, you can specify `self` to access those images for which you have launch permissions or `all` to specify AMIs with public launch permissions.

A particularly useful variant for finding an image to get started with is:

```
ec2-describe-images -o amazon
```

Examples

```
# SHOW ALL OWNER IMAGES
$ ec2-describe-images
IMAGE  ami-f822a39b  myami/myami.manifest.xml  999999999999  available  private
zz95xy  i386  machine    aki-a71cf9ce    ari-a51cf9cc

# SHOW IMAGES FOR A SPECIFIC USER
$ ec2-describe-images -o 063491364108
IMAGE  ami-48de3b21  level22-ec2-images-64/ubuntu-7.10-gutsy-base-
64-20071203a.manifest.xml  063491364108  available  public      x86_64  machine
IMAGE  ami-dd22c7b4  level22-ec2-images-64/ubuntu-7.10-gutsy-base-
64-20071227a.manifest.xml  063491364108  available  public      x86_64  machine
```

ec2-describe-instances

`ec2-describe-instances [instanceid1 [...instanceidN]]`

Lists the information associated with the specified instances. If no specific instance is specified, shows information on all instances associated with the account.

Examples

```
# SHOW ALL INSTANCES
$ ec2-describe-instances
RESERVATION  r-3d01de54  999999999999  default
INSTANCE  i-b1a21bd8  ami-1fd73376     pending  0  m1.small  2008-10-
22T16:10:38+0000  us-east-1a  aki-a72cf9ce  ari-a52cf9cc
RESERVATION  r-3d01cc99  999999999999  default
INSTANCE  i-ccdd1b22  ami-1fd73376     pending  0  m1.small  2008-10-
22T16:10:38+0000  us-east-1a  aki-a72cf9ce  ari-a52cf9cc


# SHOW A SPECIFIC INSTANCE
$ ec2-describe-instances i-b1a21bd8
RESERVATION  r-3d01de54  999999999999  default
INSTANCE  i-b1a21bd8  ami-1fd73376     pending  0  m1.small  2008-10-
22T16:10:38+0000  us-east-1a  aki-a72cf9ce  ari-a52cf9cc
```

ec2-describe-keypairs

ec2-describe-keypairs [*keypairid1* [*...keypairidN*]]

Lists the information associated with the specified key pair. If no specific key pair is given, it will list all keys you own.

Example

```
$ ec2-describe-keypairs
KEYPAIR  georgekey 98:21:ff:2a:6b:35:71:6e:1f:36:d9:f2:2f:d7:aa:e4:14:bb:1d:1a
```

ec2-describe-regions

ec2-describe-regions [*region1* [*...regionN*]]

Lists the information associated with the specified region. If no region is specified, shows all regions.

Example

```
$ ec2-describe-regions
REGION  eu-west-1  eu-west-1.ec2.amazonaws.com
REGION  us-east-1  us-east-1.ec2.amazonaws.com
```

ec2-describe-snapshots

ec2-describe-snapshots [*snapshotid1* [*...snapshotidN*]]

Lists the information associated with the specified snapshot. If no snapshot is specified, shows all snapshots for the account.

Example

```
$ ec2-describe-snapshots
SNAPSHOT snap-a5d8ef77  vol-12345678  pending  2008-12-20T20:47:23+0000 50%
```

ec2-describe-volumes

ec2-describe-volumes [*volumeid1* [...*volumeidN*]]

Lists the information associated with the specified volume. If no volume is specified, shows all volumes for the account.

Example

```
$ ec2-describe-volumes
VOLUME vol-81aeb37f  5  snapa5d8ef77  us-east-1a  in-use 2008-12-17T22:36:00+0000
ATTACHMENT vol-81aeb37f i-12b3ff6a /dev/sdf attached 2008-12-17T22:36:00+0000
```

ec2-detach-volume

ec2-detach-volume *volumeid* [-i *instanceid*] [-d *device*] --force

Detaches the specified volume from the instance to which it is currently attached. You should make sure you have unmounted the filesystem from the instance to which the volume is attached before you detach it; otherwise, your data will likely be corrupted. If the detach fails, you can try again with the --force option to force it to occur.

Example

```
$ ec2-detach-volume
ATTACHMENT vol-81aeb37f i-12b3ff6a /dev/sdf detaching 2008-12-17T22:36:00+0000
```

ec2-disassociate-address

ec2-disassociate-address *idaddress*

Disassociates the specified elastic IP address from any instance with which it might currently be associated.

Example

```
$ ec2-disassociate-address 67.202.55.255
ADDRESS 67.202.55.255
```

ec2-get-console-output

ec2-get-console-output *instanceid* [-r]

Displays the console output from the startup of the instance. With the -r (raw) option, the output will be displayed without any special formatting.

Example

```
$ ec2-get-console-output i-b1a21bd8
i-b1a21bd8
2008-12-23T20:03:07+0000
Linux version 2.6.21.7-2.fc8xen (mockbuild@xenbuilder1.fedora.redhat.com) (gcc
version 4.1.2 20070925 (Red Hat 4.1.2-33)) #1 SMP Fri Feb 15 12:39:36 EST 2008
BIOS-provided physical RAM map:
sanitize start
sanitize bail 0
copy_e820_map() start: 0000000000000000 size: 000000006ac00000 end:
000000006ac00000 type: 1
 Xen: 0000000000000000 - 000000006ac00000 (usable)
980MB HIGHMEM available.
727MB LOWMEM available.
NX (Execute Disable) protection: active
Zone PFN ranges:
  DMA             0 ->   186366
  Normal     186366 ->   186366
  HighMem    186366 ->   437248
early_node_map[1] active PFN ranges
    0:        0 ->   437248
ACPI in unprivileged domain disabled
Detected 2600.043 MHz processor.
Built 1 zonelists.  Total pages: 433833
Kernel command line:  root=/dev/sda1 ro 4
Enabling fast FPU save and restore... done.
Enabling unmasked SIMD FPU exception support... done.
Initializing CPU#0
CPU 0 irqstacks, hard=c136c000 soft=c134c000
PID hash table entries: 4096 (order: 12, 16384 bytes)
Xen reported: 2600.000 MHz processor.
Console: colour dummy device 80x25
Dentry cache hash table entries: 131072 (order: 7, 524288 bytes)
Inode-cache hash table entries: 65536 (order: 6, 262144 bytes)
Software IO TLB disabled
vmalloc area: ee000000-f4ffe000, maxmem 2d7fe000
Memory: 1711020k/1748992k available (2071k kernel code, 28636k reserved, 1080k
data, 188k init, 1003528k highmem)
```

ec2-get-password

ec2-get-password *instanceid* -k *keypair*

For Windows instances only. Provides the administrator password from a launched Windows instance based on the key pair used to launch the instance. There is no SOAP version of this command.

Example

```
$ ec2-get-password i-b1a21bd8 -k georgekey
sZn7h4Dp8
```

ec2-modify-image-attribute

```
ec2-modify-image-attribute imageid -l -a value
ec2-modify-image-attribute imageid -l -r value
ec2-modify-image-attribute imageid -p productcode [-p productcode]
```

Modifies an attribute for an image. The -l option specifies launch permission attributes, whereas the -p option specifies product codes.

Examples

```
# Add access
$ ec2-modify-image-attribute ami-f822a39b -l -a 123456789
launchPermission ami-f822a39b ADD userId 123456789

# Remove access
$ ec2-modify-image-attribute ami-f822a39b -l -r 123456789
launchPermission ami-f822a39b REMOVE userId 123456789

# Add product code
$ ec2-modify-image-attribute ami-f822a39b -p crm114
productCodes ami-f822a39b   productCode crm114
```

ec2-reboot-instances

```
ec2-reboot-instances instanceid1 [...instanceidN]
```

Reboots the instances specified on the command line. There is no display for this command, except when it causes errors.

ec2-release-address

```
ec2-release-address ipaddress
```

Releases an address that is currently allocated to you. Once you execute this command, you cannot get back the released address.

Example

```
$ ec2-release-address 67.202.55.255
ADDRESS 67.202.55.255
```

ec2-register

```
ec2-register s3manifest
```

Registers the machine image whose manifest file is at the specified location.

Example

```
$ ec2-register myami/myami.manifest.xml
IMAGE ami-f822a39b
```

ec2-reset-image-attribute

`ec2-reset-image-attribute` *`imageid`* `-l`

Resets a launch permission image attribute for the specified machine image.

Example

```
$ ec2-reset-image-attribute ami-f822a39b -l
launchPermission ami-f822a39b RESET
```

ec2-revoke

`ec2-revoke` *`groupname`* `[-P` *`protocol`*`] (-p` *`portrange`* `| -t` *`icmptypecode`*`)`
`[-u` *`sourceuser ...`*`] [-o` *`sourcegroup ...`*`] [-s` *`sourceaddress`*`]`

Revokes a prior authorization from the specified security group. The options represent the options used when you created the permission using *ec2-authorize*. See *ec2-authorize* for more details.

Example

```
$ ec2-revoke -P tcp -p 80 -s 0.0.0.0/0
GROUP   mydmz
PERMISSION    mydmz    ALLOWS    tcp    80    80    FROM    CIDR    0.0.0.0/0
```

ec2-run-instances

`ec2-run-instances` *`imageid`* `[-n` *`count`*`] [-g` *`groupname1`* `[... -g` *`groupnameN`*`]]`
`[-k` *`keypair`*`] -d` *`customdata`* `| -f` *`customfile`*`] [ -t` *`type`*`] [ -z` *`zone`*`]`
`[ --kernel` *`kernelid`*`] [ --ramdisk` *`ramdiskid`*`] [ -B` *`devicemapping`*`]`

Attempts to launch one or more EC2 instances based on the AMI and options specified. The options are:

`-B` *`devicemapping`*

Defines how block devices are exposed to the instances being launched. You can specify a number of different virtual names:

- `ami`: the root filesystem device as seen by the instance
- `root`: the root filesystem device as seen by the kernel
- `swap`: the swap device as seen by the instance
- `ephemeral`*`N`*: the Nth ephemeral store

`-d` *`customdata`*

Data to be made available to your instance at runtime. If you need to specify a lot of data, specify the data in a file and use the `-f` option.

`-f` *`customfile`*

The name of a file with runtime data to be made available to your instances post launch.

-g *groupname*

The name of the security group whose rules govern the launched instance(s). You may specify multiple security groups. If you specify multiple security groups, access to the instance is governed by the union of the permissions associated with those groups.

-k *keypair*

The public key for EC2 to place on launched instances at boot.

--kernel *kernelid*

The kernel ID with which to launch the instances.

-n *count*

The minimum number of instances to launch with this command. If EC2 cannot minimally launch the number of instances specified, it will not launch any at all.

--ramdisk *ramdiskid*

The RAM disk ID with which to launch the instance.

-t *type*

The Amazon instance type that defines the CPU, RAM, and other attributes of the instance(s) being launched. As of the writing of this book, valid values are: `m1.small`, `m1.large`, `m1.xlarge`, `c1.medium`, and `c1.xlarge`.

-z *zone*

The availability zone into which the instance(s) will be launched. If no availability zone is specified, then the instances will launch into the availability zone EC2 determines to be the best at the time of launch.

Examples

```
# Launch exactly 1 instance anywhere
$ ec2-run-instances ami-f822a39b
RESERVATION  r-a882e29b7  999999999999  default
INSTANCE i-b1a21bd8 ami-f822a39b pending 0 m1.small 2008-12-23T21:37:13+0000 us-
east-1c

# Launch at least 2 instances in us-east-1b
$ ec2-run-instances ami-f822a39b -n 2 -z us-east-1b
RESERVATION  r-ac82e29b8  999999999999  default
INSTANCE i-b1a21be9 ami-f822a39b pending 0 m1.small 2008-12-23T21:37:13+0000 us-
east-1b
INSTANCE i-b1a21bf0 ami-f822a39b pending 0 m1.small 2008-12-23T21:37:13+0000 us-
east-1b

# Launch exactly 1 instance with the specified keypair in the myapp group
$ ec2-run-instances ami-f822a39b -g myapp -k georgekey
RESERVATION  r-a882e29b7  999999999999  default
INSTANCE i-b1a21bd8 ami-f822a39b pending georgekey 0 m1.small 2008-12-
23T21:37:13+0000 us-east-1c
```

ec2-terminate-instances

`ec2-terminate-instances`

Terminates the specified instance or instances.

Example

```
$ ec2-terminate-instances i-b1a21bd8
INSTANCE i-b1a21bd8  running  shutting-down
```

Amazon EC2 Tips

We have talked about a number of concepts in this book that left open the question of how you actually implement those concepts. In this section, I attempt to put together a few recipes to help you set up and manage your EC2 environments. These tips do not represent the only way to accomplish any of the tasks they support, so there may be alternatives that better fit your needs.

Filesystem Encryption

I have recommended the encryption of your Amazon filesystems. Before you decide to encrypt, you need to balance security needs with filesystem performance. An encrypted filesystem will always be slower than one that is not encrypted. How much slower depends on which underlying filesystem you are using and whether you are leveraging a RAID. I generally use XFS on an encrypted RAID0.

To leverage this tip, you will need to have the *cryptsetup* package installed. If you want XFS support, you will also need *xfsprogs*. Under Debian, you need to execute the following as root:

```
apt-get install -y cryptsetup
apt-get install -y xfsprogs
echo sha256 >> /etc/modules
echo dm_crypt >> /etc/modules
```

The following Unix script at launch will set up an encrypted XFS volume for the ephemeral volume on an Amazon `m1.small` instance.

```
# enStratus passes in an encryption key via a web service at startup
# You can pull the encryption key from startup parameters or, for the
# ephemeral store, you can even generate it on-demand as long as you
# don't expect the need to support rebooting.
# At any rate, the key is temporarily placed in /var/tmp/keyfile
# or wherever you like

KEYFILE=/var/tmp/keyfile
# Pass in the desired filesystem; if not supported, fall back to ext3
FS=${1}
if [ ! -x /sbin/mkfs.${FS} ] ; then
  FS=ext3
  if [ ! -x /sbin/mkfs.${FS} ] ; then
```

```
      echo "Unable to identify a filesystem, aborting..."
      exit 9
    fi
  fi
  echo "Using ${FS} as the filesystem... "

  if [ -f ${KFILE} ] ; then
    if [[ -x /sbin/cryptsetup || -x /usr/sbin/cryptsetup ]]; then
      # Unmount the /mnt mount that is generally pre-mounted with instances
      sudo umount /mnt
      # Setup encryption on the target device (in this case, /dev/sda2)
      sudo cryptsetup -q luksFormat --cipher aes-cbc-essiv:sha256 /dev/sda2 ${KFILE}
      if [ $? != 0 ] ; then
        echo "luksFormat failed with exit code $?"
        exit 10
      fi
      # Open the device for use by the system
      sudo cryptsetup --key-file ${KFILE} -q luksOpen /dev/sda2 essda2
      if [ $? != 0 ] ; then
        echo "luksOpen failed with exit code $?"
        exit 11
      fi
      # Format the filesystem using the filesystem specified earlier
      sudo mkfs.${FS} /dev/mapper/essda2
      if [ $? != 0 ] ; then
        echo "mkfs failed with exit code $?"
        exit 12
      fi
      # Remove the current entry in /etc/fstab for /dev/sda2
      sudo perl -i -ane 'print unless /sda2/' /etc/fstab
      # Create a new entry in /etc/fstab (no auto mount!)
      echo "/dev/mapper/essda2 /mnt ${FS} noauto 0 0" | sudo tee -a /etc/fstab
      # Mount the drive
      sudo mount /mnt
      if [ $? != 0 ] ; then
        echo "Remount of /mnt failed with exit code $?"
        exit 13
      fi
    fi
  fi
```

You will need to leave the key file in place unencrypted under this model. enStratus allows you to delete this file after the script executes because it will pass the key in as needed.

The one weakness with any of these approaches is that an unencrypted version of the key is stored on the unencrypted root partition, even if for a short time under the enStratus model. One trick to get around this issue is to do the following:

1. Auto-generate a random key and store it in */var/tmp*.
2. Mount the ephemeral store in */mnt* as an encrypted device.
3. Erase the auto-generated key.
4. Pass in a permanent key from enStratus (or other tool).

5. Store that in */mnt/tmp*.
6. Mount an EBS volume using the permanent encryption key.
7. Delete the permanent encryption key.
8. Use the EBS volume for all critical data.

With this approach, no critical data is hosted on a volume whose key is ever unencrypted, except at the local operating system level. The storage of the permanent key is on an encrypted volume that uses a temporary encryption key.

If you really want to go to town on the encryption front, fill up the drive by raw-writing random data to the drive before putting it into use.

Setting Up RAID for Multiple EBS Volumes

I have seen huge boosts in disk I/O performance by tying several EBS volumes into a RAID0 (striping) array. On the flip side, I would not expect to see tremendous redundancy benefits from RAID1 (mirroring), since all EBS volumes must be in the same availability zone as the instance to which they are attached. Unfortunately, my experiences are in line with Amazon expectations, which have shown that mirroring of EBS volumes erases most performance gains of striping, so RAID5 and RAID10 end up being pointless.

To be fair, I have not done extensive benchmarking of all the options using a variety of filesystems. I generally stick with a single encrypted EBS volume on each instance, except when performance is critical. When performance is critical, I will join the volumes into a RAID0 array. I may or may not encrypt that RAID0 array, depending on the balance between security and disk performance.

When designing your RAID0 setup, keep in mind that the parallel writes to the disk that make RAID0 so fast have an upper limit at the data exchange rate that connects your instance to your EBS volumes. The optimal number of drives and sizes of those drives will therefore depend on the nature of your application.

I use the *mdadm* package to set up my RAIDs:

```
sudo apt-get install -y mdadm
```

The following script will create a RAID0 with two drives:

```
SVCMOUNT=/mnt/svc
# Pass in the desired filesystem; if not supported, fall back to ext3
FS=${1}
if [ ! -x /sbin/mkfs.${FS} ] ; then
  FS=ext3
  if [ ! -x /sbin/mkfs.${FS} ] ; then
    echo "Unable to identify a filesystem, aborting..."
    exit 9
  fi
fi
echo "Using ${FS} as the filesystem... "
```

```
ISNEW=${2}

ARGS=($@)
DEVICELIST=${ARGS[@]:2}
DEVARR=($DEVICELIST)
DEVCOUNT=${#DEVARR[*]}

if [ ! -d ${SVCMOUNT} ] ; then
  sudo mkdir ${SVCMOUNT}
  sudo chmod 775 ${SVCMOUNT}
fi

# Verify you have devices to mount
if [ ${DEVCOUNT} -gt 0 ] ; then
  # If more than one, set up a RAID0
  if [ ${DEVCOUNT} -gt 1 ] ; then
    map=""
    for d in ${DEVICELIST}; do
      map="${map} /dev/${d}"
    done
    # Create the RAID0 device s /dev/md0
    yes | sudo mdadm --create /dev/md0 --level 0 --metadata=1.1 --raid-devices ${DEVCOUNT} $map
    if [ $? != 0 ] ; then
      exit 20
    fi
    # Configure the RAID in case of reboot
    echo "DEVICE ${DEVICELIST}" | sudo tee /etc/mdadm.conf
    sudo mdadm --detail --scan | sudo tee -a /etc/mdadm.conf
    # Are these newly created volumes or remounts of old ones/snapshots?
    if [ ${ISNEW} == "true" ] ; then
      # Make the filesystem
      sudo mkfs.${FS} /dev/md0
      if [ $? != 0 ] ; then
        exit 24
      fi
      echo "/dev/md0 ${SVCMOUNT} ${FS} noatime 0 0" | sudo tee -a /etc/fstab
    else
      echo "/dev/md0 ${SVCMOUNT} ${FS} noatime 0 0" | sudo tee -a /etc/fstab
    fi
  else
    # Just one volume, not a RAID
    if [ ${ISNEW} == "true" ] ; then
      # New volume, create the filesystem
      sudo mkfs.${FS} /dev/${DEVICELIST}
      if [ $? != 0 ] ; then
        exit 29
      fi
      echo "/dev/${DEVICELIST} ${SVCMOUNT} ${FS} noauto 0 0" | sudo tee -a /etc/fstab
    else
      echo "/dev/${DEVICELIST} ${SVCMOUNT} ${FS} noauto 0 0" | sudo tee -a /etc/fstab
    fi
  fi
  sudo mount ${SVCMOUNT}
fi
```

APPENDIX B

GoGrid

Randy Bias

THERE IS MORE THAN ONE WAY TO BUILD A CLOUD INFRASTRUCTURE. Amazon Web Services (AWS) represents the "service infrastructure" approach, offering customers a number of customized, nonstandard, but highly scalable services to rebuild an infrastructure in a totally virtual environment. GoGrid represents more of an industry-standard approach that offers a very familiar data center-like environment, but in the cloud.

For many customers, the GoGrid approach is easier and more comfortable because it reuses traditional idioms and technology, such as VLANs (virtual LANs), network blocks, hardware appliances (load balancer and firewall), and file storage (SAN or NAS, storage area networks and network attached storage, respectively). This approach represents what we call a *cloudcenter*: a data center in the clouds.

Types of Clouds

Infrastructure clouds (aka "Infrastructure-as-a-Service," or IaaS) can be built primarily in two ways: service infrastructures or cloudcenters. Both allow all of the capabilities one expects from IaaS:

- Scale on demand
- Pay-as-you-go

- Conversion of capital expenditures (CapEx) to operational expenditures (OpEx)
- Programmatic (API) and graphical user interfaces (GUI)
- Basic infrastructure: storage, servers, network, power, and cooling

Although both provide the same basic value, these two approaches differ significantly in approach:

Service infrastructures
: This is the approach made familiar by AWS, and described in much of this book. Service infrastructures are essentially custom web services "in the cloud." These can be used individually or composited together to deliver a web application or do batch processing. For example, Amazon offers servers, storage, databases, queuing/messaging, payment processing, and more. Every one of these web services is a unique and custom solution. Storage using S3 uses the S3 protocol and storage mechanisms. The AWS SQS queuing service uses its own nonstandard custom protocol and message format. The same goes for SimpleDB, their database service. These services were designed in a custom manner to allow Amazon to scale to 50,000+ servers and thousands of products. They are being repurposed as publicly consumable web services that AWS customers consume for their own uses within their business models.

Cloudcenters
: Most AWS competitors use this approach. Its methodology is to provide standard data center services using standard technology and protocols, but in the cloud. Storage is available via familiar protocols, such as SMB/CIFS and NFS. Databases are provided using standard SQL and RDBMS. Firewalls and load balancers are based on hardware appliances instead of custom distributed and configured firewall software.

At the end of the day, the choice is between a custom infrastructure with its own protocols and standards you'll need to conform to, or a more traditional data center-like service that has the same benefits, but is delivered using industry standards.

Cloudcenters in Detail

GoGrid is the first and largest U.S. cloudcenter and is popularizing this approach. (Other cloudcenter companies include FlexiScale, ElasticHosts, and AppNexus.) Among its primary advantages is the ability to directly translate skillsets, existing infrastructure, and projects to the more flexible cloud environment. GoGrid's approach will also eventually make so-called "cloud-bridging"—connecting and integrating your internal data center to external clouds—much easier.

Data Centers in the Clouds

Traditional data centers are composed of the following elements:

- Perimeter security using a hardware firewall and intrusion detection system
- Load balancing using a hardware load balancer
- Network segmentation using differing network blocks and VLANs
- A combination of physical hardware and virtual guest operating systems
- Filesharing using (NAS)
- Block storage using (SANs)
- Data center support services: DNS, DHCP, server imaging, inventory management, asset management, and monitoring
- Power, cooling, bandwidth, and backup for all of these support services
- 24/7 on-site support and staff

Cloudcenters are very similar to traditional data centers, offering most of these services with only small variations to provide them in a multitenant fashion. In addition, cloudcenters, unlike normal data centers, deliver direct cost efficiencies along with indirect human efficiencies through the GUI and API.

GoGrid Versus Traditional Data Centers

The primary downside of traditional data centers is the need to build them out to maximum capacity. You, as the IT owner, are required to forecast correctly and to build and manage your capacity in-house.

Cloudcenters allow reusing your current in-house data center expertise with external cloud providers. Knowledge and expertise reuse means less time spent learning new paradigms and more time on driving real business objectives. Retain control while gaining the virtues of clouds in terms of OS licensing, CapEx avoidance, forecasting, and capacity management.

Additionally, all of the benefits of cloud computing are available to you, such as adding capacity on-demand, automating workload elasticity, and paying only for what you use. This allows IT staff and managers to optimize their infrastructure costs both internally and externally by moving elastic workloads to the cloud and expanding/contracting usage as necessary.

Horizontal and vertical scaling

Deploying on GoGrid is like deploying in your internal data center, but with many new tools that can accelerate and smooth your operation processes. Like any other cloud, you have the option to "scale out" (horizontal scaling). In contrast to most other clouds, you also have the option to "scale up" (vertical scaling) using not just virtualized cloud instances, but real dedicated physical hardware. This is similar to a regular data center, where there is usually a mix of virtual and physical servers. You can get direct access to physical hardware with lots of RAM, high-speed direct-attached-storage (DAS), and many dedicated cores, all on the same virtual private network.

Use the right tool for the job: virtual servers for stateless workloads (centralizing processes but storing no data) that you can scale out easily and physical servers for stateful workloads (manipulating data in a persistent manner, such as in a database) that are easier to scale up. Here is the reasoning behind the choice between these options:

Scaling out (horizontal)
: It's easiest to scale out for servers and use cases that are relatively stateless, such as web servers, application servers, and batch processing. With these kinds of workloads, adding an additional server usually requires little or no additional configuration or architecture work. Simply adding the additional servers allows you to add more capacity.

Scaling up (vertical)
: In contrast, scaling up is best for stateful applications and workloads such as databases and fileservers. In these cases, simply adding additional servers does not directly translate into more capacity. Usually you will need to do some significant reconfiguration, change your architecture, or at least automatically balance the state data across new servers. The majority of your data is in these servers, so rebalancing or synchronizing terabytes of data dynamically is a nontrivial task. This is why it is usually desirable to simply use bigger servers rather than more servers. It's also why these servers tend not to be dynamic in nature, even in clouds.

This is why GoGrid supports both dimensions of scale. It's not always sufficient to scale out, and scaling up is a viable and important scaling tactic. Amazon recognizes this, and consequently provides different server sizes in AWS. Virtualization, however, is inherently a strategy for making a single server "multitenant," meaning that it holds multiple customers or applications. If a part of your application can use all of the capacity of a modern physical server, it makes no sense to run that application in a virtual server. Virtualization in this case adds unnecessary overhead, and if you need more computing power or memory than the cloud offers, you have to twist around your architecture to scale out horizontally.

On the other hand, most applications will hit a scale-up constraint at some point, and then you will be forced to scale out. You will need to measure demand and match capacity to it in either scaling model. But with Moore's Law still in full effect, it will be a long time before physical hardware is not a valid scale-up strategy for most applications.

GoGrid deployment architectures

A typical GoGrid deployment looks like one in your own data center (Figure B-1).

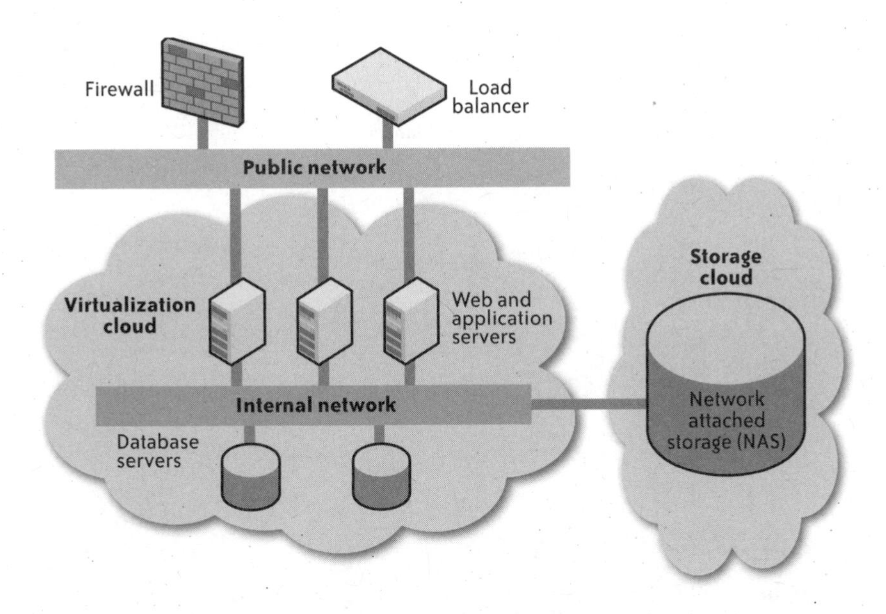

FIGURE B-1. Roles of virtualization and the cloud in GoGrid

In typical physical data centers, the application servers face the Internet to interact with the users, while the backend servers are protected by an extra DMZ-like layer and all systems securely share a NAS. As in a traditional data center, there are two network segments (VLANs), one for the public "frontend" and one for the private "backend," which uses private IP addresses (RFC1918). Just like in your data center, there is a NAS (GoGrid Cloud Storage) for use with home directories, backups, archives, and related storage needs.

Scaling up on GoGrid does not look very different from scaling out, except that the all-important high-performance databases run on dedicated physical hardware, as shown in Figure B-2.

Focus on Web Applications

The cloudcenter architecture is friendlier for web applications than for batch processing applications, which may not need all of the traditional data center infrastructure, such as firewalls, load balancers, and VLANs. Most batch-processing applications do well in utility and grid computing environments. GoGrid can also be used for batch processing, and although batch processing is an important component of many web applications, we optimize the environment for your core transactional web application, which we believe is far more important.

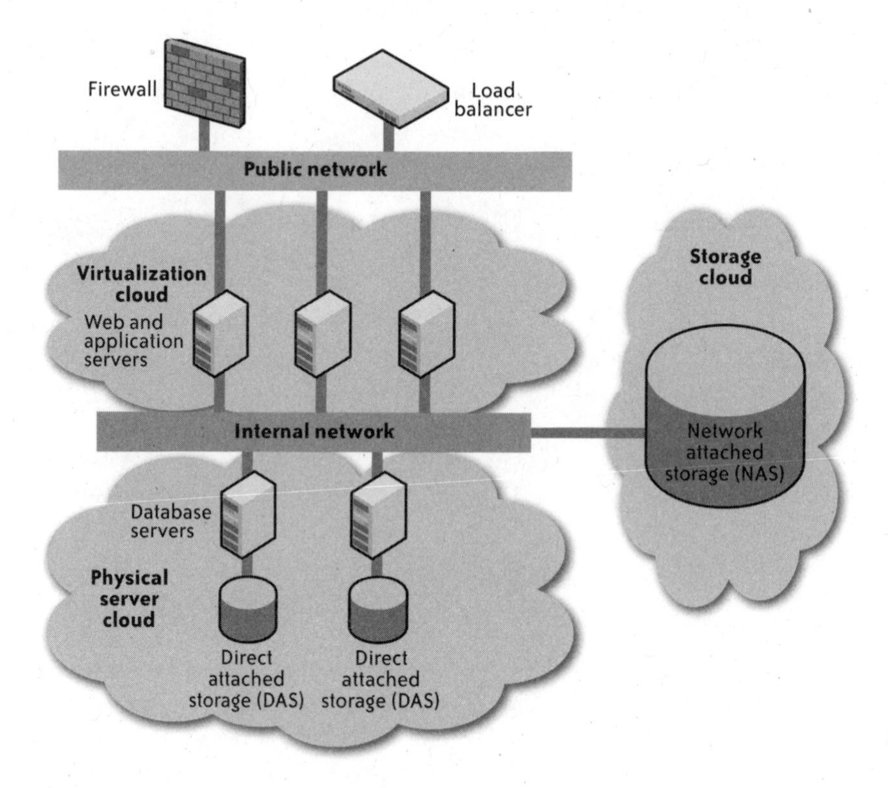

FIGURE B-2. Role of physical hosting in GoGrid

Comparing Approaches

When comparing cloudcenters (GoGrid) to service infrastructures (AWS), it's important to remember both the practices of traditional data centers and the kind of application you are deploying.

Side-by-Side Comparison

It may help to look at traditional data centers, cloudcenters, and service infrastructures side-by-side. Table B-1 lists some of the areas of functionality for each type of infrastructure and how they compare.

TABLE B-1. Functions and capabilities of clouds

Functionality	Traditional data center	GoGrid (cloudcenter)	Amazon (service infrastructure)
Firewall	Perimeter hardware firewall	Perimeter hardware firewall (Q1 2009 release)	Custom distributed software firewall
Load balancer	Hardware load balancer	Hardware load balancer	Roll-your-own software load balancer (possible 2009 release of custom load balancer service)
Network isolation	VLAN	VLAN	Faux "VLAN" separation using distributed software firewall
Private networks	Yes (VLAN)	Yes (VLAN)	No
Network protocols	No limitations	No limitations	Restricted; no multicast, no broadcast, GRE and related may not work
OS choices	Unlimited	Some limits	Some limits
DNS	Yes; managed in-house	Yes; managed by GoGrid	No
Persistent local storage	Yes	Yes	No
Persistent network storage	Yes	Yes	Yes
Mixed virtual and physical servers	Yes	Yes	No

As you can see, cloudcenter-style cloud architectures are very similar to traditional data centers.

Real-Life Usage

The differences between cloudcenters and service infrastructures will become apparent the minute you attempt to try both GoGrid and AWS.

With AWS (service infrastructure model), little of your current in-house expertise in networking or storage is relevant. You will need to learn new skills to manage S3 and even extend your server system administration skills to include managing EC2's additional server paradigms, such as runtime metadata, the lack of multicast and broadcast network traffic, server groups, and their custom distributed software firewall.

In contrast, GoGrid's approach (cloudcenter model) is very similar to using the console of VMware VirtualCenter or another virtualization management system. In addition to servers, you can control your network, DNS, storage, load balancers, and soon firewalls through the same integrated UI.*

With either system, you'll be able to get the standard cloud computing benefits of scaling on demand and paying as you go.

What's Right for You?

Ultimately, the right choice for you will depend on your application, business needs, and the traditional "build versus buy" factors. When considering the buy option and comparing that with cloud computing solutions, you will evaluate whether it's important to focus your efforts on adapting to the custom idioms of different service infrastructures, or if it's better to use a cloud that looks more like what you are used to, but in the cloud.

Randy Bias is VP technology strategy of GoGrid, a division of ServePath. He is a recognized expert in the field of clouds and data center infrastructure, with more than 20 years of experience with ISPs, hosting providers, and large-scale infrastructure. He writes frequently on infrastructure and cloud computing at his personal blog: http://neotactics.com/blog.

* The GoGrid release of hardware firewall support is slated for 2009.

APPENDIX C

Rackspace

Eric Johnson

RACKSPACE IS PROBABLY THE BEST-KNOWN COMPANY when it comes to traditional hosting services for corporations. Its roots lie in managed hosting of physical servers and its trademarked Fanatical Support, allowing most customers to rely on its technology experts for stability and performance. Rackspace's customers range from one-server configurations to complex configurations spanning hundreds of servers and high-performance network gear.

But Rackspace is fully aware of the trend in this industry, as demonstrated by Amazon.com's EC2 and S3, to move away (at least partially) from dedicated physical servers into the "cloud," where some customers are comfortable running in a shared environment.

Rackspace seeks to provide a full suite of options for customers, ranging from dedicated physical servers to complete virtual servers, along with "hybrid" environments that are a mix of the two.

Rackspace's Cloud Services

Rackspace's Cloud Division is the umbrella group that builds and delivers Rackspace's core cloud technologies to customers. By combining product offerings built at Rackspace, a few key company acquisitions, and its existing managed hosting offerings, Rackspace intends to provide a comprehensive offering. Customers can range from small startups in a complete virtual/cloud environment up to complex physical servers, and anywhere in between.

Many current Rackspace customers have begun to see advantages to using cloud services. By developing and offering cloud services within Rackspace, customers have the advantage of a single vendor relationship, fully integrated technologies, and, in many cases, performance benefits from having their physical servers located within the same data centers as their cloud services.

Cloud Servers

In October 2008, Rackspace acquired Slicehost (*http://www.slicehost.com*), a leader in the Linux virtual server hosting market. Mosso, a subsidiary of Rackspace, will be leveraging this technology to develop Cloud Servers, an offering similar to Amazon's EC2.

As outlined in this book, an attractive use of EC2 is to move your dedicated physical servers into the cloud and replace them with virtual EC2 instances. Arguably, EC2 is more attractive to customers that need temporary computing power or the ability to grow and shrink their configuration as needs demand. Slicehost, in contrast, was built specifically for customers looking at less-expensive dedicated hosting, and provides true virtual server hosting. The Slicehost offering is designed specifically for dedicated 24/7/365 hosting and utilizes host servers with RAID(1+0), redundant power supplies, Internet backbone providers, etc.

Over the coming year, the core Slicehost technology will be combined with other features to provide customers with dynamic abilities that are similar to those described in this book for Amazon's EC2. Slicehost as a company will remain a standalone entity and continue to improve and expand its product portfolio. The Cloud Servers product built with Slicehost technology will consist of:

- Custom images and image repositories
- Programmer APIs
- Static IP addresses
- Utility-based pricing
- Microsoft Windows support
- The traditional Rackspace 24/7 Fanatical Support

Cloud Servers will also serve as a dynamic resource that Rackspace's managed hosting customers can tap into when they need extra computing power. This "hybrid" capability will provide the best of both worlds and allow customers to temporarily spin-up virtual nodes for data processing or for "burst" demand.

One major way Rackspace customers can benefit from Cloud Servers will be to spin-up environments that duplicate their physical configuration almost instantaneously. These additional environments can be used for testing, QA, development, change reviews, etc. They can be brought up for a limited time and then powered down again until the next need arises.

Cloud Files

Cloud Files is a storage service very similar to Amazon's S3 system, described earlier in this book. In early 2008, Cloud Files began a private Beta, and in October 2008 it entered into full public Beta. Like S3, Cloud Files is not intended as a direct replacement for traditional storage solutions, but offers the following features:

- Organize data (Objects) into storage compartments called Containers, which are nonnested (i.e., nonhierarchical).
- Objects can range in size from zero bytes up to 5 GB.
- Custom metadata can be associated with Objects.
- Customers can create an unlimited number of Objects and Containers.
- All features and functionality are accessible through a web interface and programmer APIs (ReST web service, PHP, Python, Ruby, Java, C#/.NET).
- Services can be utilized on a pay-as-you-go basis, eliminating the need for over-buying/under-utilizing storage space.

A good use-case for this type of storage system is for storage web content (images, video, JavaScript, CSS, etc.). Cloud Files has the added benefit of publishing that content behind a proven, industry-leading Content Distribution Network via Limelight Networks' infrastructure. This truly makes CDN easy for the novice and affordable for everyone. Users simply need to create a Container, upload their data, mark the Container "public," and combine the Container's CDN-URI with the Object name to serve it over the CDN.

Cloud Files also shines as an unlimited data store for backups and data archives. You may be one of the few users that performs full backups of your personal computer, but chances are you do that to an external drive or DVD media. That backup data is physically near the computer and will be lost in the event of a fire or other site disaster. Backing up to Cloud Files ensures that you have cheap "infinite" storage that's accessible anywhere you have an Internet connection.

Like Amazon's S3, Cloud Files is "developer friendly." It provides a RESTful web services interface with APIs for several popular languages. The data stored in the system (Objects) can include custom metadata, which is commonly used to set user-defined attributes on those Objects. For example, uploaded photos could include metadata about which photo album they belong to.

Cloud Sites

Although application hosting is not covered in this book, a summary of Rackspace's cloud offerings is not complete without some mention of Cloud Sites. Originally developed by Mosso, a subsidiary Rackspace group, the original product offering has evolved into an automatically scalable web platform hosting solution in a LAMP (Linux, Apache, MySQL, PHP/Perl/Python) stack, Microsoft Windows (.NET, ASP, SQL Server) stack, or both.

Amazon's EC2, Slicehost, and Cloud Servers require customers to do all their own server and application configuration and management. If all you plan to do is run a website in a traditional web stack, Cloud Sites handles the backend management so that you need only focus on your site. The Cloud Sites technology even automatically scales the site(s) as demand increases or decreases.

Fully Integrated, Backed by Fanatical Support

In 2009, these discrete services will be fully integrated to provide a robust cloud offering to Rackspace customers. They will have the option of mixing physical and virtual servers, "infinite" storage, and fast CDN-enabled websites that automatically scale with demand.

Furthermore, at the end of the day, Rackspace is a service company, employing expert technology professionals to provide Fanatical Support for all of these new technologies.

Eric Johnson, or "E. J." as he is better known, is the software/product development manager for Rackspace's Cloud Files storage system. His after-hours job is working with and managing the senior engineers within Racklabs, Rackspace's Research and Development group.

For the last 15 years, E. J. has worked solely with open source technologies in various roles, including Unix system administration, networking, DBA, and software development. He has performed these functions in the airline and aerospace industries. Over the years, he has contributed his time to the open source community by supplying patches for SSH, being a package maintainer for Arch Linux, and authoring technical guides on DNS/Bind.

E. J. holds a B.S. in electrical engineering from Drexel University in Philadelphia, Pennsylvania, and an M.S. in computer science from Rensselaer Polytechnic Institute in Troy, New York. He currently lives in San Antonio, Texas, with his wife, Julie, son Tate, and a second son due in April 2009.

INDEX

Numbers

A

B

C

We'd like to hear your suggestions for improving our indexes. Send email to *index@oreilly.com*.

G

H

I

J

K

L

M

N

O

P

R

S

T

U

V

W

X

Z

ABOUT THE AUTHOR

George Reese is the founder of two Minneapolis-based companies, enStratus Networks LLC (maker of high-end cloud infrastructure management tools) and Valtira LLC (maker of the Valtira Online Marketing Platform). During the past 15 years, George has authored a number of technology books, including O'Reilly's *MySQL Pocket Reference, Database Programming with JDBC and Java,* and *Java Database Best Practices.*

Throughout the Internet era, George has spent his career building enterprise tools for developers and delivering solutions to the marketing domain. He was an influential force in the evolution of online gaming through the creation of a number of Open Source MUD libraries, and he created the first JDBC driver in 1996—the Open Source mSQL-JDBC. Most recently, George has been involved in the development of systems to support the deployment of transactional web applications in the cloud.

George holds a B.A. in philosophy from Bates College in Lewiston, Maine, and an MBA from the Kellogg School of Management in Evanston, Illinois. He currently lives in Minnesota with his wife Monique and his daughters Kyra and Lindsey.

COLOPHON

The cover image is from *http://www.veer.com*. The cover font is Adobe ITC Garamond. The text font is Linotype Birka; the heading font is Adobe Myriad Condensed; and the code font is LucasFont's TheSansMonoCondensed.

Related Titles from O'Reilly

Web Authoring and Design

ActionScript 3.0 Cookbook
Ajax Hacks
Ambient Findability
Creating Web Sites: The Missing Manual
CSS Cookbook, *2nd Edition*
CSS Pocket Reference, *2nd Edition*
CSS: The Definitive Guide, *3rd Edition*
CSS: The Missing Manual
Dreamweaver 8: Design and Construction
Dreamweaver 8: The Missing Manual
Dynamic HTML: The Definitive Reference, *3rd Edition*
Essential ActionScript 3.0
Flex 8 Cookbook
Flash 8: Projects for Learning Animation and Interactivity
Flash 8: The Missing manual
Flash 9 Design: Motion Graphics for Animation & User Interfaces
Flash Hacks
Head First HTML with CSS & XHTML
Head Rush Ajax
Head First Web Design
High Performance Web Sites
HTML & XHTML: The Definitive Guide, *6th Edition*
HTML & XHTML Pocket Reference, *3rd Edition*
Information Architecture for the World Wide Web, *3rd Edition*
Information Dashboard Design
JavaScript: The Definitive Guide, *5th Edition*
JavaScript & DHTML Cookbook, *2nd Edition*
Learning ActionScript 3.0
Learning JavaScript
Learning Web Design, *3rd Edition*
PHP Hacks
Programming Collective Intelligence
Programming Flex 2
Web Design in a Nutshell, *3rd Edition*
Web Site Measurement Hacks